Destination: Me
108 Ascetic Days across Eurasia
by Martins Ate

2010

It is too little to show gratitude to my parents: mom Marite who gave to me all her beauty of youth sharp mind and courage; dad Ilmars, who gave everything that his kids are prepared for the life, well-educated and mindful thanks to what I have found peace in various situations of a life including this trip.

The Love of my parents, support and their care is the biggest treasure I have ever got.

It is thanks to the good heart of my parents that this book has made its day out – so that also they can see the places I have carried them in my heart.

Earth provides enough to satisfy every man's needs but not every body's greed.

Gandhi

Sometimes it's not about the choice of being lonely. It's a choice of being lonely -WHERE. Sometimes even in most crowded place we are lonely but sometimes when we are alone can feel fulfilled.

I just hope that during this trip I will feel fulfillment. Now I'm already in the train to Brussels Charleroi Airport where I take a flight to Poland, Katowice and then hitchhiking till Kiev where I have a first stop to take a look at the beautiful monasteries and orthodox churches built even thousands of years ago.

Hope to reach my destination by tomorrow's evening. Then will head towards Moscow. Usually when I'm leaving I feel liberated. Now is different. It feels like I'm ... Just leaving... I even did not packed my bag until the last minute where I dropped in everything I was using for last few days.

I'm leaving unprepared. And I'm already out of Brussels and there is a 60 euros in my pocket! That's all I'm taking with me.

I prepared my visas before the tip to countries where it could be hard to obtain in on the way - like Russia, China and India. Actually it would be useful if I would have managed to get Kazakhstan visa as well but I didn't.

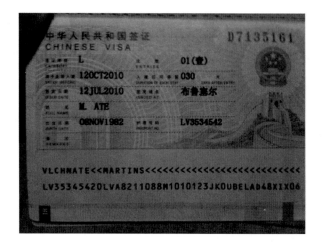

I took a train from home which delivered me until the airport in Brussels. From there on I went into the wild. Out of my comfort zone and without any even expectations on what's next.

My way from Katowice Airport and Ukraine border was nice. I stopped a lorry and it dropped me almost till the Ukraine border after I was picked up by two other family cars.

Pleasant talks in a car about my trip, e.t.c. When I reached the Kiev I felt dead, but a first thing I did was went for bath in a cold water and listened to the female voices of morning prayers at Feofania monastery. Feofania Monastery is some place which leaves a long time impression!

When I entered the Ukraine, I decided to take a small nap on the lake during the daytime. The place I found was a great lake and yammi berries all around it. Few horses and fisherman also was around.

Went to some monasteries, visited places I used to go every day while lived there just a year ago. My mentality is more Slavic. I lived there for 2 years and probably would live there even now if because of work I wouldn't move to London.

After visiting other monastery in Kiev, called Pecherska Lavra I'll be heading north to Moscow.

I reached Russia and so far that felt great I would say. As i'm hitchhiking all this trip it is very interesting to see a different people and different lives.

Yesterday, when was in Kiev, my heart was full of pain. Somehow I have not forgot the moments I had in that beautiful city.

When I entered Russia on the border between Ukraine and Russia on the border, before my passport was given back to me, the border control officer said to me: " you will be beaten until all your bones will hurt, you will be punished with no feelings, and you will be sorry to come to Russia... IF you will stay in Russia longer than your visa expires."
So I have only about three weeks to cross the Russian - Chinese border from where I plan to visit Chinese sacred mountains after seeing friend from Beijing.

While we got closer to Moscow, the car broke down - something went wrong with electricity and we stopped. When went out of the lorry - we saw that batteries located underneath the car has somehow felt of and got broken. But Denis - the Lorry driver - said - now worries - Russian Technic will work anyway. He fixed some cables, stopped another lorry which helped to push the lorry with a rope and after 3 minutes we already continued moving to Moscow! Amazing.

I was very tired. Very. And decided to stay here instead of moving forward to Moscow. And the lorry driver Denis decided to show me the Russian Welcome and showed the local places.

He explained me that it is tradition to have a Vodka as I'm first time in Russia and during our trip around the Kaluga City he introduced me to his friends.

So we drove to his friends, took some girls and went to the lake Andreyevskiy, had some bear. Water was as a fresh milk - so hot!!!!!!! But Russian girls are hotter!!!!!!! !!And Russian heart - the hottest!

So we took a couple liters of Vodka and moved to the home where Denis lives. During our way from the store to home we dropped by some more girls and had a smoke of local ganja I was already drunk so did not felt any changes... So we drunk a Vodka with some local chiks and I noticed that I love it. I love the Russian warm welcome, this anarchy that is in the air, all thees churches, true people. This truly is the amazing and great country and it is impossible even to think about understanding of Russia. Just enjoy.

Kaluga. There is a saying - you can not understand Russia using your mind. The same happened to me yesterday. I was thinking - how possible is this - to be responsible for the largest country in the world - how is it - for a president to rule the country as big as eleven time zones?! It is amazing. So big and huge country!!! Sooo big!

So we moved on further travel together after a heavy night of party and drinking in Kaluga to Tatarstan. About 300 km before Ural mountains which divides Europe from Asia about 1500 km from Moscow land is seeded with oil pumps. I was still with my new Russian friends and we have been driving already over 12 hours since this morning and are thinking of getting some rest in the nearest forest with water so can swim and make up some fire to have a bit of a warm food.

On our way the car broke down during our way because of a heat - but Denis very quickly jumped out of his seat, opened the hood from which smoke was coming out, made few things, filled something with water, inserted one rubble somewhere where electricity is and in 15 minutes we continued our trip. Russian technique...

Then during our way we saw how local gangsters where trying to hook up some lorry from Rostov... Russia ... very original country...

Since I left Kaluga I have not seen any cars with EU number-plates. I'm thinking about going to Kazakhstan and from there to China - just I don't know do I need visa and how is the road to China and from China's Kazakhstan's part to India - as I might imagine that there could be some mountains on my way for which I'm not prepared. Also visa to Kazahstan now would be useful. Actually I just now realized that actually I'm not prepared for this trip at all - just recently, for example I noticed that my tent is for a beach - it's without a floor ... And I don't have a sleeping bag...

The next worrying thing is about registration thing in Russia - there is a law which says I need to register in some local immigration service otherwise when leaving country I might be arrested and get some rest in Russian prison. Hope it won't affect me.

I guess I smoked to much herbs instead of preparing myself for this trip. I would need to have at least a tent and a sleeping bag. Well - somehow I'm sure I will get some pot here to smoke my worries away...

Samara. There is a problem with wi-fi travelling like this. I'm already driving more towards unknown part of so massive Russian country. Current position is about 1000 km from Moscow and I'm heading towards Cheljabinsk.

I'm still with the driver I met couple of days ago while hitchhiking from Kiev to Moscow and who showed me the Russians warm heart, welcomed me with vodka and Russian beauties, who gave me a roof for one night, shower, nice company and food. But for me the most important thing during this trip is to cross the Russia in 20 days as my visa expires on 20th of August and today is already 25th July.

Ural mountains. See you, Europe, good morning Asia! Serpentine of mountains covered with pines, lot of small rivers with weird names. One o'clock in a night. Tiredness makes us search for a place to stay for a night. High in a sky except the light coming from the full moon, on the east side brightly shines the planet of wisdom - Jupiter. The air is so fresh that I can feel it's sparkles even through the closed windows of the car.

Somewhere in Northern part of the Ural mountains where gravity law is working in opposite direction - when you drive up the mountain the car goes easier up than it's coming down and when you drive downwards you really need to push on a gas pedal.

We decided to go as far as we can to escape tomorrow's mountain movement of heavy loaded lorry and trucks who have stopped all among the snaky mountain road.

Feels really cool to cross the Europe and Asia border by hitchhiking! The worrying part is my "diary's" battery life and absence of internet. I've already forgotten and dropped out of my mind to have a possibility of Wi-Fi access when we finally would have a stop for a short break in Ural mountains around 2 am; I quickly made a fire - Probably the quickest fire in my life - straight away when stepped out of the car took some wood, walked ten meters ahead, carefully placed them in scouts style and in half minute the flames where almost until my belly. We had a quick noodle soups with a water boiled in Russian army soldier's cattle.

After we continued driving and on our way up by the serpentine, just about hour later the second time our car broke down. The car just stop driving and we had to drive of the road. It was a midnight and finally Denis found out that the gas pedal has broken down. Well - for cars made in Russia it is usually and Denis easily fixed the car in 5 minutes using his knife and screwdriver.

The Sunrise is already making dark blue sky turn into a light purple and ... Denis has been sitting by the steering wheel already 22 hours and we are high in a mountains with a car who is breaking down each 1500 km and it's driver each time fixed any problem even without turning of the engine. Any comets?

Wonderland....

As the average speed is about 70-80 km/h the road goes slowly and we have a plenty of time to talk. We talk about religion, God, knowledge and history. How Russia was built, how mankind survives through all changes and listening to such a Russian modern classic as singer Coi, Krutoi, Viagra and Nautilus Pampilus.

We are now getting closer to the river Volga witch for Russians is as sacred as for Hindus Ganga and will definitely take a break to swim in there. My new friends are heading towards

Kazakhstan as Denis has born there, but his friend accompanies him because both of them got the vacation since Friday.

So far God has been mercy-full to me and has helped me to eat, drink, sleep and move towards China by gallops. Just with Denis I've been driving now already about 2000 km. It helps me a lot.

Yesterday, before leaving Kaluga, based 180 km from Moscow, we went to the local churches and placed some candles for the health of our beloved ones, parents and for the fortunate trip. I was only 60 km from Pustinskovo Monastery where I planned to take some photos and videos of morning prayers, but as I was vodking all previous night decided to do it may be on my way back - probably I will fly from Australia back to Russia to finish things I needed to do - still have left some monasteries to visit. But I'm not planning anything ahead so let's see by situation.

The truth still stays the same - I was not preparing for this trip at all except taking my passport to different embassies to get a visas and creation of this website which I did while was on high from my regular Brussels - Amsterdam bong trips and 3 gr. smuggle to stay on high longer.

We plan to be in Chelyabinsk by tomorrow - so from there the new challenge will turn it's face to me as I'm gonna be on my own again. In the third part of this amazing country covered by forests, lakes, rivers, steps and deserts where I will be in about a week in the region of Amur.

There is 10 rubles in my pocket left which is less than one euro. And the deeper we go the less trees are around. We are getting closer to step. It's the hottest summer in Russia and there is crisis situation in 23 regions - all corn and wheat is dying out because of there has not rained more than two months.

Everybody I've met strongly believes in God, walks to the churches, and gives donations and prayers in there. But everybody drinks vodka the same time. It is amazingly how such a things can find a harmony in daily Russian life and create such a great nation. Amazing. I can't stop wondering about it.

After we spent a night at the Ural Mountains by some lake with a crystal clear water, we moved towards Chelljabinsk which now was only about 100 km from us. I spent morning swimming and washing myself up in a lake.

On our way we stopped at the road cafe (rigalnik) to have breakfast and tea with lemon. I used my ginger which I took from Brussels to help my stomach get rid of all the unformulated meals it had since I left home.

While we had a breakfast local TV channel at the cafeteria displayed weather news which made me a bit worried. The cities I'm heading towards to - like Omsk, Kemerovo, Novosibirsk - temperature is almost twice less than I expected: only + 12; + 14 degrees Celsius during the daytime which means that I might experience weather I'm not prepared for at all. The hot Moscow's weather is gone. And I found out that my tent actually has a floor and it made me so happy.

The Sun is not so hot anymore behind the Ural Mountains and everybody can feel it. Instead of yesterday's +37 we have only + 24 and it's already 4:30 PM.

My new friends are very keen to me and we have a lots of a great chat, but this is the last day with them together as after about 500-600 kilometers I'm moving towards North. The round-trip might take for me at least two days as I need to go by the country roads about 400 km until next big city - Omsk, 650 km from Novosibirsk.

Currently I'm already in the region of Siberia. I feel amazed and a bit scared from the possibility of my body to resist the cold for next few weeks. I have only one pair of jeans with me. But mostly I feel alright and until now I haven't regretted I moved towards this unknown experience. I pray. I guess only God can sort my trip in most attractive, interesting and less damaging way.

Outside is so cold that people are wearing winter jackets kind of I'm usually wearing from November till March. But I feel excellent because I've reached Omsk - a city I have heard a lot about before.

Yesterday's Vladimir, who picked me up from Kazakhstan border and drive even further than he needed to and gave some number to call while in Omsk made me a pleasant surprise.

After walking all night until finally some truck driver picked me up, and driving with him all day until we reached Omsk, I called to the number I got yesterday from Vladimir and as told introduced myself as Martin from Belgium. The pleasant woman's voice on the other side of the line explained that they have been waiting for me and gave me the address.

I looked in a map of the Omsk, found a street and in 3 hours reached the destination on my way walking into a very beautiful Russian Orthodox Church.

When I entered the colorfully decorated building and introduced myself - some nice young lady with a pretty smile brought me downstairs to Russian sauna, where she, unfortunately left me be on my own, where I finally chilled for a few hours and swam in a pool after 120 degrees Celsius hot black Russian sauna.

Tomorrow the owner will take me to the immigration service and to one local new built cathedral in city of Omsk, from where I will start moving towards Novosibirsk.

That's it - I'm done with Omsk and now heading towards Novosibirsk. Thank you, wonderland and its inhabitants!

It's midnight again, middle of Russia, middle of summer and southern part of Siberia. Now I'm moving towards Kemerovo from where towards Irkutsk, at the lake Baikal. It's so cold outside that I'm afraid of going sleep and I can see the air coming out from my nostrils.

I'm in a middle of a forest and surrounded by weird sounds. As it's Russia it's OK to expect some wolfs and the Russian national symbol - brown bear. So my attention stay's on high especially if I'm in Novosibirsk, Siberia.

Few hours later I finally decided to stop for the night just before turning to Kemerovo because it's almost 3 AM and nobody can see me on a road.

I walked inside a crop field and was hurry to make up my tent as local Siberian mosquitoes where attacking me even worse than a frost - while making up my tent I got about 50 bites only on my head and hands.

Now I'm "safe" if this word can be used at all inside the tent and putting on all my clothes as it's incredible cold. Probably outside is about 4 to 5 degrees Celsius.

Here where I've stopped is a biggest possibility to catch some lorry straight away to Irkutsk - which is only now about 1800 km. I'm planning to stay at the lake Baikal for a few nights then continue my journey towards China.

But for now - I've put on ALL my clothes, even swimming shirts as it's still very, very cold. Maybe not as much cold as my unpreparedness for this trip makes me feel about it. I need to get some sleeping bag for the next night like this or start searching for a warmer place to stay overnight already during daytime.

In the morning I waked up from cold. It was so cold that all my clothes didn't helped out much. Through the battle with mosquitoes at 6 AM I collected my tent and moved forward towards Kemerovo and Irkutsk.

Но если есть в кормане
Пачка сигарет
Значит все не так уж плохо
На севодняшньый день (Цой)

Translation:
But if there is in the pocket
Pack of cigarettes
It means it's not so bad
For current day (Tsoi)

Basically - never is so bad that couldn't be worse! i need to change my strategy. I will put forbid for me to sleep during the nights because of a cold as if temperature gets a little more down I might get frost. But this is the official last week of the summer in this region - so it will get colder and colder with each day.

Just about a hundred kilometers up North asphalted road ends. The last big city is Tomsk ad after that only some small villages. It's about -45 degrees Celsius in the wintertime in here and sometimes people are heating up their houses all year long. Just a two hours driving towards North from here starts tundra and wetlands.

Just this morning I realized that: I AM IN SIBERIA with a summer clothes.

I'm heading from Western Siberia towards Eastern Siberia and people here are uniquely kind, welcoming and helping. Kemerovo is the city of coal miners. I got here with 4 cars, totally around 300 km from Novosibirsk and it's a midday.

I was driving with one Russian army guy who gave me a gift of an army food pack for extreme situations, the other guy gave me a Russian flag, but the third - Edik from Kemerovo drove me all the way from the city to the auto route to Irkutsk, which is about 1800 km from here and insisted to take 200 Russian rubles from him, which I did as he was telling me that he is giving it from pure heart and I'm totally out of cash right now. I feel real gratitude towards all Russia and I'm truly amazed about their wide, warm heart.

So now I'm heading towards Marinsk, from where to Irkutsk. Every step I make I can feel - I'm in Siberia. It is as it sounds. Rough, cold weather but air full of Love. It is something western civilization might never understand and get the feeling of.

Towards Krasnoyarsk I was picked up by funeral car, who drove me about 600 kilometers with 5* service as there was even a plasma screen with DVD and all the way long I was watching Russian movies. On our way we stopped in some road cafeteria where I had some proper warm meal. What a coincidence it was when other expedition from Netherlands also stopped there - they are driving with quadrocycles all around the Russia, Kazakhstan, Turkey, etc. We had some chat and afterwards we continued our funeral car journey.

Actually it was very weird day since the very beginning as the first driver who picked me up was from local veterinary service, second from army and third from funeral service. I stepped out of the car in Krasnojarsk at midnight, again and as it was to cold for sleeping without sleeping bag I decided to continue my journey.

As I continued walking during the midnight somehow I felt scared for the first time during my trip. There was something in the air what made me walking instead of hitchhiking all night.

First thing was that each time I was preparing to raise my hand, my shoe ties opened. After it happened three times I decided to look upon it as on the sign that probably it's not rather good place or time for hitchhiking and continued just walking.

My fear did not left me for hours especially after seeing Police lights getting closer to me. I decided that probably something has happened somewhere around the area I was I hided myself from the sight of the road.

I was very tired and eyes where closing even if I was walking. I felt kind of paranoia obsession covering me. I felt fear from everything. Totally I hided myself from around 200 cars instead of hitchhiking.

After analyzing reasons for my paranoitical behavior I decided that probably my fear comes because of funeral car. The second reason was that probably really something was happening around me and I did right that I was a bit more auspicious than usually. Anyway I decided to chant some mantras to put my wondering mind back on stripe and it turned my mind back to it's original state - peaceful, fearless and assured and started hitchhiking again but as it was already around 3 AM there was no any cars.

So by continuing walking I made about 25 Km only walking. By back was hurting because of the heavy bag I'm taking with me and finally around 7 AM I was picked up by some local oil magnate who was driving brand new Lexus Hybrid SUV.

He dropped around 150 KM towards Irkutsk which is located at the Baikal Lake, where I'm going to stay few nights. But still it's more than a one and a half thousand kilometers to go on the almost off-road roads.

So I was picked up by lorry driver Kostya who will drive me till the very Irkutsk and I was so happy for such a luck as I won't need to talk the same story to more people - where, from where, why and how I decided to do this trip.

Roads from Krasnoyarsk are really bad. It is true off-road. We have to make around 700 Km when the road got so bad that our driving speed is just a little bit faster than walking. I'm knocking out for half hour after each half hour until finally I have enough energy to write this.

It's evening already and we damaged our tire so I helped to change it. We did it quite quickly and currently are crossing river Uda which seems like a bit overflowed.

Probably we might get till Irkutsk until tomorrow's morning and I really hope that on the day 11 I will be hanging around lake Baikal carelessly for at least few nights. I really need to recharge my batteries, have some rest and peace before going to the "next level adventure" - Hitchhiking in China.

We are listening to Russian 80' and 90' hits. Excellent! wouldn't even imagine better music for this trip! that's it. I'm knocking out again while we are crossing off-road Russia with a huge, fully loaded truck over 25 tons of weight with speed of bicycle.

As we stopped last night 220 Km before Irkutsk, after waking up in the morning we started replace the tire. I helped with opening and closing of the bolts, the same as yesterday only today we needed even to put a tire on a disk as yesterday we already used the spare wheel and today we had only a tire with no wheel which was locked into a cargo section on top of the boxes with watermelons.

Again, one of our tires got bad. It's all-ready one o'clock past midnight and we decided to take a nap until tomorrow's morning and after replacement of it continue our journey.

As I'm lighter than the driver, I was climbing on top of the boxes to get it. It took around an hour and afterwards we continued our journey.

Weather is nice and it seems like after cold Baikal region night it's gonna be a hot Siberian day as already at 10 AM is so hot that both windows are opened and we are moving with a speed of around 70 km/h as the road seems to be better than yesterday.

We are passing one of Siberian Salt mines. The day is getting shorter and shorter as we are moving towards East and it is already 8 hours different between Brussels and here.

I've been driving now with this driver already 24 hours and his wife is calling him at least three times a day, asking him how he is and telling that she can't wait for him to come home even if they are together for 15 years now. That makes a man want to come home.

The man itself by the nature is made to work, care and take responsibility about his family but there must be something which drives him to do it - and that is a woman's love and care towards him: the feeling of home - when he gets there; the feeling of peace. Woman needs a home for comfort, but for the man it's a place to rest and restore the energy he loses in work - to give the woman that comfort she needs.

And we are moving more towards South the weather gets warmer. In Novosibirsk, from where I'm driving now to Irkutsk the temperature raises up from 16 degrees Celsius during the daytime in Novosibirsk till 30 in Irkutsk. That makes me happier!

So now I'm in Irkutsk some Japanese restaurant where is a free wi-fi and tee Afterwards I'm going to move towards Baikal lake and get some rest for a couple of days.

In Irkutsk I was dropped at the District of Lenin, from where I got on the buss number 6 and drove till the very center of the city for only 12 rubles which is about 25 Euro cents.

So after hopelessly searching it around the city Irkutsk for one more hour I decided to move towards direction of Chita - out of the city. i walked around the kilometer after I noticed that finally my Wi-Fi indicator becomes alive and I found that his life is coming from some Japanese restaurant and I moved towards it.

The door was opened for me by some Asian girl who greeted me with a low gesture. I explained her that I'm an tourist and I'm not hungry but I'm here just because of internet and all I need, probably, is a tea. She seated me and brought me a tea with a maximum kindness I even did not felt to be earned to receive and after asking her how much dos a tea costs she replied that it's free as it is usually a complimentary drink and it's free even if I'm here only for a tea. I felt uncomfortable regarding this free tea, kind service and myself - brutal, tired, probably a bit stinky and restless tourist who cared only about the internet.

So I took a seat, plugged in charger my iPhone, chatted with friends, replied to emails and updated website for around two hours and enjoyed a beautifully served green tea for free and when I moved away, I left 50 out of my 65 rubles to her as a tip - as anyway it was cheaper and more, much more pleasant experience as if I would spent it in internet cafe.

Afterwards I continued moving towards the South exit of the city and saw few churches on my way I heard about so decided to pop in at the first of them.

Together I visited three local churches, SvjatoTrojeckij, Church in the name of the Saviour Nerukotvornovo Obraza and the central one and afterwards continued my road outside of the city towards Lake Baikal.

I was picked up by two 30 years old woman and a man who was a husband of one of them. It was a very pleasant and attractive company who stopped after each 25 kilometers to have some vodka and I was unable to refuse it... They dropped me till the very lake Baikal and told to come to pick me up tomorrow for a collecting of strawberries.

On the Baikal are two types of winds: Sarma - the cold one and Arguzin - the warm one. Even if it seems like today is Arguzin - I have all my clothes on me again and I'm burning fire from a local woods at the very coast of this incredible lake.

The temperature is about +5 degrees Celsius. And as i'm making the fire i've prepared to spend here all night. After the fire was gone I took the stones and covering them under the sand, put the tent upon it so I kept warm all night.

Last night i made up my tent at the shore of the lake Baikal and burned a fire. The weather was quite cold so I used the stones from the fire place and was putting them beneath the tent so I could sleep on them as they where warm.

It is incredible feeling that you can swim and drink in the same time as it is the cleanest lake in the Russia. There are 330 rivers flowing inside it and only one out of it. But when the morning came - my Russian friends came as well by screaming: Latvian, where are you (in Russian) and I waked up.

They had a vodka with them and a food so afterwards we moved to their garden where I was collecting strawberries and drinking - that's right - vodka, again. So after a short brake with a vodka and strawberries I'm back on stripe again and heading towards Chita, which is about 1000 Km from here and then to China.

The car who picked me up was from the local winter sports base and when the woman found out that I have come along all the way from Europe, she was so glad that she showed me her small and cozy camping and as I was very drunk, she told me to stay for a few hours, get some rest and made me a bed.

After a short day-brake I continued my trip and was picked up till Ulan Ude city by some youngsters who drove me about 350 Km. From there I walked around 20 km and visited sacred mountains of shamanism, where they even nowadays kill and burn animals to get a blessings from gods. In this part of Russia Buddhism and shamanism are main religions.

From there towards Chita was a real car deficit and I was picked up by a great guy Konstantin, we even drove till the healthy Radon lake on Russian army truck which he bought 5 days ago. And now I'm walking to Chita where I hope to get the wi-fi connection and move towards Beijing. This must be my last stop in Russia. Chita - the city with a lot of Chinese, only about 700 Km from China.

When I came till the border of Russia and China on Zabaikalsk and Manchurian the first thing I found out that I cannot cross the border walking. When I started catch the cars to drop me on the other side, one lady picked me up and after I explained my situation she was laughing on me and told that it will be very funny when in china will born my children with Chinese faces and with a blue eyes as due to my trip plans this is the best I can expect.

The truth is - it's forbidden to hitchhike in China. And more - I need to have a person who is watching me all the time which is usually provided in tourism offices. And if I will be cached I will be definitely moved to Chinese prison with fee of around 40,000 Yuan's a d if I will not pay them I will be moved, after spending some time in prison, to some country side with obligation to work until I pay them off. And as the salary in China is low - I will work until I'll find a wife and even will raise up a children.

That did not stopped me. Even if the story was quite scary but did made me think seriously for the first time during this trip. I asked more people afterwards on the border and all of them said the same thing and even worse. One truck driver told me that in the Northern part of the China people are living really poor and even a kids has got a knifes. They are killing even for a pair of jeans. The truck driver told that only in the last year three of his colleagues where stubbed by Chinese kids in the back in Manjuria and he really advise me not to go further.

The fact is that Manchurian is kind of a trading point between Russia and China and people going deeper into China must have a agent. But I don't have a money for it. So I went to the local pawn shop to sell my golden chain with the amulet which has been made especially for me according to my date, time and place of birth but the price they gave me still was not enough to continue my journey so I kept the gold.

I was so upset. All my trip has stocked because of some Chinese laws. So I moved through the city towards back to Moscow. A lot of youngsters in Zabaikalsk came to me and asked for stupid things just to beat me up around the corner and take my belongings. But after I explained that I don't have a money and cell phone they left me alone. I've hidden my phone inside a bag, wrapped in papers and socks. But the money - I truly have got only 150 rubles which is about 3 Eur...

It's a step out here: no place to run and no place to hide. Even after exiting the city someone was still following me with a car everywhere I went and time to time coming closer and driving around me like a flying bugs. I just slowly moved towards the highway. After walking like that around two hours the rain came together with a sunset and my followers disappeared.

I walked about half hour more inside a step where no trees to hide - just mountains and grass are and quickly made up a tent as I decided that it is too late and I'm to tired for any decisions regarding my trip and it will be the best to leave it for tomorrow.

As I was all wet and the tent was wet as well all my clothes did not helped me much of being attacked by the cold nights of Russia. But somehow I managed to get my night-sleep. Thank's God i'm alive and I have all my belongings with me.

I waked up together with a sunrise. It was raining all night and as it was very cold I was waking up about 10 times to change my position and warm myself a bit up with a pranayama exercises. But as I was sleepy and tired it was hard to do it for long enough to feel warm enough for a well night sleep.

In the morning the problem stayed the same - is this a finish of my trip? Do I need to turn back and hitchhike all the way back to Brussels? Or should I just go, just do it, whatever the consequences might be.

I cannot go back. I need to finish my journey as planned, even if it takes a Chinese prison. Fuck that. I need to make it. I need to do this whatever it takes.

My feet are burning due to everyday long distance walks. I'm in a bad mood thinking that now I need to cross all Russia again. I cannot go back. No. Not this time.

I made up a tent and chanting prayers moved back to the border. While changing I had an idea of going by train but again - how will I make it without money at least till Beijing where is more civilized and safe for me to continue the trip and where I could meet my friend Ingus, who lives around there for a few years now.

So I stopped by the train station and was waiting for him by his doors after found out from the guards that he will be around 10 AM. As it was early morning, I waited for a few hours and was planning to introduce myself with my ideas and probably get some free pass towards Beijing by train as I was thinking: "what is a one passenger driving for free for such a great, big and financially stable company as the Russian Railways.

When the Director came, I introduced myself as Martin from Latvia and asked him for a time to have some conversation with me. After finishing few urgent things, he invited me inside his office.

The party begins. This is a moment where it is decided not IF but WHEN I'm going to be in Chinese prison by hitchhiking - before or after Beijing.

Alexander - the managing director of the Russian Railways Zabaikalsk Station was very strong standing man with a strong handshake. I liked him from the very beginning. He made an impression of a good businessman. Behind his chair was a photography of the President of the Russia - Medvedev. Also, in his huge office was a flags of Russia and China. He was polite, strict and tolerate.

After I introduced myself and briefly told about the project "vedatrac" where I'm planning to go from Europe till Australia hitchhiking with a total of 53,000 Km, I explained the reason of me coming to him as the only hope

Alexander listened to me carefully, made few calls to confirm the seriousness of my situation and asked me a simple question: what do I want from him. My answer was as short: ticket till Beijing. He smiled and picked up a calculator.

After pressing some buttons on it he took his wallet, pulled out of it three thousands of rubles and told that he is unable to provide me with a free ticket to Beijing but from Manchurian the trains are cheaper and this amount should be enough to get me till there by Chinese trains. He told me that He is unable to get for me a free ticket. Gave me the three thousand rubles and told: "Мы же Русские!" (We are Russians!). I could not stop thanking him and his words where sounding in my ears long after I left him. I guess they will remain in my memory for a long time as it is incredible how Russians are helping me out all the way long through the huge Russian empire. Alexander saved me. Saved my trip. Saved my dream. I left his office with a greeting from him to catch up one day when I will be a millionaire and will do a business together. This is what I call a True Russian. With a heart full of love and understanding. Thank You, Alexander.

China border crossing, Manzhouli - I'm in China. Thanks to Vladimir! to cross the border I needed to pay 300 rubles to local bus service. They provided me with excellent service and assisted me on every step during the passport control and customs.

Entering the Manzhouli is a real culture shock. From naked step with even no trees, old houses to the big city, full of gardens and huge houses, filled crowded with people. Just ten kilometers away from Russia and it feels like I'm on another planet.

I went to the local bank and exchanged rubles to yuan's and moved towards railway station after my ticket to Beijing with eyes wide open. Back to civilization... welcome.

I bought a ticket in a last minute and my train towards Beijing, where I'm going to drop off in Tianjin was departing in 10 minutes. I paid less almost on half than expected - instead of 500 yuan's I paid only 219 yuan's but I had a seat instead of sleeping place. I just don't know how long does it takes till get there and I hope that somehow I'm not will be able to contact Ingus when I'll come there.

The train is quite comfortable and I'm leaving deeper and deeper into the China and I'm the only European inside this whole train. All eyes on me!

On my way towards Beijing I found that I'm not the only "white" one. I met two girls from Irkutsk, where I was just less than a week ago. They are driving not the first time and explained me a lot about local traditions and life in China itself. They gave me even a cellphone to contact my friend in Tianjin.

So "I'm cool" now and my trip in China is going much smoother than planed and everything is even much better than it seemed yesterday.

The train is crowded with local Chinese who spit on a floor a d everybody is eating sunflower seeds. I'm going to be in Tianjin in about 30 hours. But the worst thing is that when I was buying a ticket - there was only standing places available - that means that as soon as someone new comes in who have booked a seat is throwing me out of it. And currently I've been standing now already for 5 hours. And I have some 20 more to stand. It is actually quite absurd.

As I found out later, the VIP standing seats are sleeping in a sink, which is based at the end of each doorway beside the toilets. Pity this is my first time in China and I didn't knew it earlier...

On the seventeenth day I woke up only about midday still feel sleepy. Ingus introduced me to a lot of people already who is learning Thai-boxing, kung-fu and karate in here meanwhile working as an English teachers.

My visa in China expires in about a month and I think I'm going to stay here for about a few days or even a week to find out properly what to do next and where to go forward. I really like in here. It is a great country.

It's so good in here that actually I don't want to go any further. The Chinese girls are better than I expected, too. So a lot of things might come up in here!

Next day I got up at 2 PM because of yesterday's going out to some club and drinking beer, we decided to finally go for a walk to see at least something more from Tianjin than just a beer and weed.

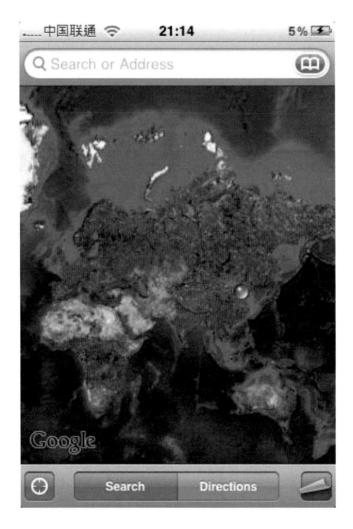

Everything is crowded and it seems like there is no rules on the roads and traffic is moving as it happens, with no rules at all.

I like it here and more and more with every day I'm willing to stay here just need to decide when - should I continue my trip and come back afterwards or stay here now.

So it's been a third week of my travel, and I'm still in Tianjin. Yesterday I visited some factory of electric ... Some things ... I'm not quite sure about... But it seemed to me like something very interesting.. If anybody understands it - please let me know. Probably someone needs them somewhere?

And my friends from Kaluga (close to Moscow, Russia) who I met during my trip sent me some forgotten photos with vodka and some Russian girls. It made me remember the good times I spent in Russia.

Generally I could divide Russia into three parts: the one until Ural mountains - very European part of the Russia; from Ural mountains till lake Baikal - real Russia - friendly and very helping; and Russia after lake Baikal - the part I didn't enjoyed much but there are some beautiful things to do as well.

So, I'm thinking of to take off for the rest of my trip on this weekend. I still have to go to Tibet where is needed some special permit and reach my main goal - mountain Kailash.

Soo... After long vodka tours in Russia and resting in Tianjin I feel ready to move forward and leave the cozy apartment of my friend Ingus where I stayed almost 10 days and had a great vacation in China with everything I needed to continue my ascetic expedition.

I'm planning to leave on Sunday. And now I understand that the Russia was the easiest part. Like a level I; with a level II following in China since the very beginning. I just had a good luck that I'm here in Tianjin only thanks to Alexander, Zabaikalsk train station director in Russia.

The local hitchhikers taught me everything I need to know about going inside a Tibet without a special permission. It might sound bad - but I'm going there illegal as the only way to et there legally is to buy a travel package through some travel agency who only then is legitimate to issue a travel pass to Tibet. The average of such a package costs a lot of money which I do not have..

I will use train to get until Tibet, Lhasa, from where I will need to hitchhike like invisible through the police checkpoints more than a thousand kilometers in a mountains with altitude from 3000 to 4500 meters to get till the Mt. Kailash which is one of the ten most beautiful mountains in China according to Chinese National Geography.

Actually, Mt. Kailash is considered a divine mountain universally by people from all over the world. Its shape is very much like the pyramids in Egypt with four nearly symmetrical sides. There are a few temples around it and many wide-spread legends shroud these temples with mystery.

I'm still preparing morally. To be ready to handle some situations I can't imagine now and definitely have never experienced before. I'm not scared of it.. It's more than a one way thinking of that I need to get there no matter what - so I just need to prepare and to be ready for a lot of things, such as I can be deported from China with a huge fee, I can be dropped in some prison, I can get cold or just simple starving as there are no many places to get a food as it was in Russia... So I'm preparing. Mentally. Just a few last days and then I'll be off to Lhasa and Kailash.

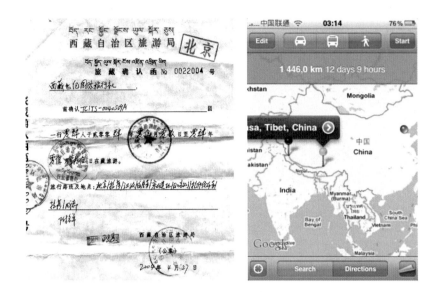

The blue dot is where I'm now and the red - Green - Lhasa: from the blue till the green I'm taking a train. But still I need to manage somehow my fake Tibetan Permit as I'll need to show it in a train towards Lhasa... So it's going to be an adventure starting on Sunday!

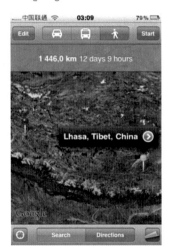

And the red one - is a city about 200 km after Mt. Kailas.. Somehow I need to get from there to India at the end. And I have only 20 days left to do this as my Chinese visa expires after that.

Yesterday my friend Ingus bought me a train ticket till Lanzhou - about 30 hour travelling through Beijing where I will be like in about 7 hours from now.

From Lanzhou I will take a train to Lhasa, Tibet and afterwards will hitchhike deeper into Tibet - towards mt. Kailash.

I think I will be out of reach for next 2 weeks... My bag is getting heavier and heavier not lighter with every day and every stop I make. In Russia it was a water who made my bag heavier but now when I'm heading towards Tibet - it's my new equipment who makes it heavyish.

Ingus gave to me a sleeping bag which can hold a temperature down to -5 degrees Celsius which I think might be useful in Tibet and tracking around the Himalayan Mountains. Also we bought a mat to sleep on in a tent to avoid cold from the ground.

And - I'm thinking of getting rid of something in my bag - I just don't know in what I will be in a need during my trip. So I'm keeping e everything at least until I'll reach Kailash.

The real reason why I'm driving now to Lanzhou is because I've heard that from there the chance of getting to Tibet train is higher than from Beijing where Tibet Travel Permit control is much higher.

So I stepped out in Beijing and had a small walk around the city for about three hours. The city truly impressed me. So big. The buildings are huge. And everything is developing, so many new buildings and development structures.

In Beijing also I continued moving towards Lhasa by train to Lanzhou. The train was so crowded that few people even could not get onto it and stayed on the platform.

Somehow I got luckily in but - forget about a seating place - there is no even a place to put my bag and I had to wait for about 10 minutes until finally I found some place for it.

The train till Lanzhou goes for about 20 hours and besides the fact that there is no space even to stand, the salesman are going around the train all the time selling food and drinks. I'm allocated by Chinese people who are trying to speak with me and asking a lot of different questions about Europe and are taking pictures with me. I'm a star on this trains! (Everybody gets its five minutes of fame once in a lifetime) - I guess these are mine!

After spent night at the train where I could not resist willing to sleep and knocked out for a few hours under the bench on the floor, I'm finally in Lanzhou.

(My jumper on a Chinese princess)

It feels like the real reason population growth fast in China is because of trains. Everybody at the end is so friendly as close as the canned fishes. Everybody is sleeping on each other shoulder, sharing food, clothes and even siting places goes around the clock in a line so everyone at the end gets at least half hour of a seat.

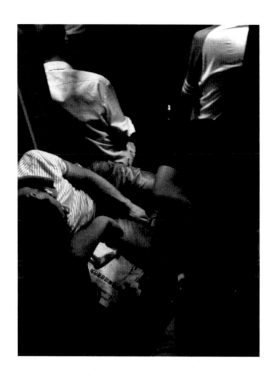

I was not alone who was sleeping under the bench. My idea seemed to be good for Chinese princess who joined me at around 3 am under the same bench. We talked before already I just did not expected that she might get with me under the bench.. We shared same bag as a pillow and warmed up each other by laying side-by-side. It was quite romantic experience, but I was knocking out as I was not sleeping last night, too. So we just where laying side by side until the morning when she prepared a breakfast for me, too, from the food she had with her in a train. I think in longer journeys it's easy to find a Chinese wife and get married straight on a train!:D

Train stopped at Lanzhou. Intuitively I got the point of the city even without stepping out of the train - similarity with Russian Chita is obvious - lot of hustlers and youngsters looking for they daily pray - tourists.

I wiped away bad thoughts by chanting mantras and as soon as I left a train station I noticed a Buddhist monk in front of me who was wondering around the same way I would do if I wouldn't met him. I came to him and after failure of getting conversation in English I told him straight: "Lhasa?". He smiled and I followed him towards ticket boxes.

He was going to Lhasa as well. With assistance off dictionary I explained that I need a help of buying a ticket to Lhasa as I'm driving there without a travel permit which is required for getting a ticket. My friend in Tianjin explained that the only way of getting a ticket to Lhasa from here is by asking someone. I felt so lucky that the one I'm asking to buy a ticket for me to Lhasa is nothing else than a monk from Lhasa. That felt right!

Train for today was already gone and we got a ticket for tomorrow - I just followed my new guide and we went to the hotel as there was more than 20 hours till next train. We both moved into one room for approximately 10 Euro for both, made some tea and dinner, I took a shower in a toilet (I mean it literally!

I got up because of the pulsing Adrenalin in my blood as I'm so stressed out about Tibet and how will I manage to get pass all the check-points unnoticed. I just can't sleep even if in last 48 hours I've slept maybe 4 to 5 hours in a train under the bench with a princess by my side.

I waked up the monk and we both went for a night-out in Lanzhou... we just walked across the city without talking as we cannot understand each other and I was wondering all the time about how he walks. As we were walking together the same road it seemed like he never got bothered by such obstacles as other people on a road moving slower than we did; barriers on a road - it seemed like he was just flying by them unnoticeable, with ease, never slowing down the same time I needed to slow down or change my track to overcame them. And he was moving much beyond the average walking speed.

It was interesting for me and I started observing him to find out his secret. I noticed that he is not even looking around and once I even stopped him just before it looked like one of the cars would hit him. We went to some outdoor cafe where we had a ice-tea and moved back to our hotel room.

On our way back I tried to catch his mind vibrations to try to get into that state when nothing is an obstacle. And I felt his mind turned towards something which is beyond streets, cars, people around, lights and advertisements. It felt like he overflies each second spent outside by guiding himself to some higher source instead of one's eyes can see. When I finally somehow made myself turn away from the flashy streets and put myself into the state that I'm walking through the streets not because I'm interested in it but because I need to get back to the hotel to rest and gain powers to move forward next day and instead of advertisements and prostitutes on the streets I started seeing My Mission of this whole trip and My Vision - it started happening to me as well and I felt like I learned a lesson. I stopped noticing crowded streets and never again had to stop or slow down my walking speed. I felt like I'm not there. I felt in a moment which is not interesting for me so by this detachment to it it detached from me and stopped bothering me at all. Even if there where people walking they where the ones who gave me a road, where stopping or walking aside.

So it reminded me how important is to "Just Do It" style of acting. Of course this is a small thing; but as the small electrons and protons are running around in the atoms; same way planets are running around in the solar system and around the all stars. My point is that same rules governs each second and it does not matter how big things are. Compering to the whole Universe our Solar system is as small as comparing to the table and chairs atom is. It's just our limited mind who puts our imagination in boundaries "big" or "small" so we could live more easily.

Today I learned to overfly the crowd. Tomorrow I can overfly checkpoints!, if the lesson is learned, I thought to myself. There are no people who get's on our way accidentally - I'm sure that it is not coincidence that I met this monk today. And it is not coincidence that I decided to leave Tianjin on the day I did; and it is definitely not a coincidence what he by coincidence showed me today.

Everything and everybody is bonded together that there is no even a second of coincidence. Never. Everything happens for a reason even if at the end everything is generally pointless. It's just the experience to gain, willing lesson to learn and move further towards our own wishes. So if there is anything that matters - it's our wishes which are meaningful to us and generally pointless..

Even this trip matters as much as it matters to me, but generally it's pointless as I just fulfill my wish to do this for a reason that matters to me. But as from each dead-end are at least two exists there must be at least two other ways to fulfill my reason for willing to do this trip. And by living through the situation in mind it's possible to get rid of the reasons which puts us into an action to get fulfilled and so get rid of all the time ongoing wishes. By-flying people on the streets > situations > wishes > actions; and taking only one thing from whole life experience - the thing that makes us grow and be capable of getting to the Truth and The Beginning of All Things. To be "back home", happy; to be as we were meant to be.

Maybe just sometimes it's just more easily to live through the wishes and situations by physically living them through. Like now my trip I'm doing - for me it is easier to get the desired results by taking this trip, but I'm sure same result I will get at the end might be achieved in other ways. And if it is so - this trip generally does not matter at all or it does matter as long as I'm willing to do it or achieve something because of it which can be achieved in at least two other ways. It's just my choice. But it doesn't mean nothing. And it is not pointless as long as I'm not willing to do it in any other way. Reason - for any action it is only one: to be happy > fulfilled > whole > in the true state of being - as a man was made to be.

So all the pointless things matters only as long as we give them a matter because of our will which we cannot control and more or less never know what we want and are searching for it by wishing more of everything in a hope it will fulfill us and make us happy.

Actually all we need is a true knowledge about what we actually need to be happy. And True Knowledge can be gained only from the Truth and One who created the Truth. Need to be "on a line" with the Creator of the Truth to get The Truth. I guess that's what yoga is about.

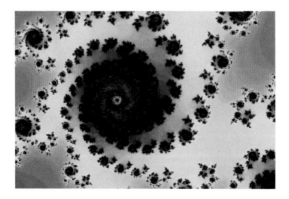

... (a bit of a philosophy) (: ...

Nothing and Everything is the same. It depends on our wishes if it's nothing that is everything or it is everything that is nothing. And our wishes depends on our consciousness. And our consciousness depends on how are we - as a human beings - doing our human being The Only True Duty - searching for the Truth.

As every animal can eat, sleep, fuck and produce babies but only human is capable to reach for the Truth it is that what human should do. So that's what should be on our mind before taking any wish into an action. Are we doing that what we were born for, are we doing our duty for gaining Understanding of things not Knowledge about them.

What did we did today that could bring our minds closer to the level where if it wishes something - action towards it - will produce a long term happiness not just a short-term satisfaction. And when will we understand these things so we could get a power to act according to our true wish - to be happy and make others surrounding us happy!

Happiness will come as a reward of doing a Duty of being a Human. Search the Truth within The Creator! That's the only Truth waiting for us to be found. That's the only happiness for a human being.

We went sleep till next day to start a forbidden part of a trip - after 250 from here is a Tibetan border where for the first time I will be asked for a permission to enter and I have not got any ideas of which lies I should use to get till Lhasa and will they help anyway...

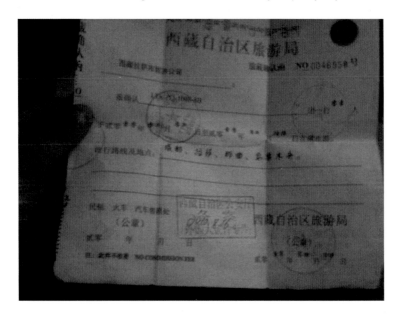

The beginning of the HARD PART

As soon as I entered train station police lady came to me and asked for a permit. I explained her that it is coming later and took a ticket out of her hands. But the permit isn't coming at all...

After waiting for a while in a waiting lounge I decided to walk around a little bit as the police lady was watching me. And went to another waiting lounge leaving my bag with a monk and I saw some "white people" on my way, who just came from Lhasa. They are some guys from Canada and are travelling around much like I do just in a legal way. They adviced me a lots of things as the girl use to work in Lhasa for a tourist company.

Word by word and I asked about the permission and explained that I don't have one. They said it's impossible without it but she remembered that they do have the old one which they used to but with the expired date and issue date: 2010.08.06 and today is a 2010.08.17. They gave that permit to me and I added a number 1 with a black pen before 06 so it looks like it was issued just yesterday!

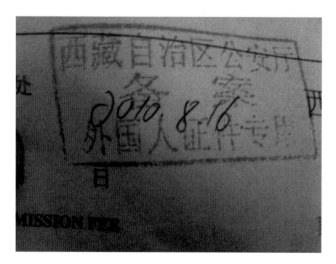

(the number "1" before "6" is my masterpiece - anyway - I don't understand a word what's written in there - just added "1" as I felt it might work - but none was sure about it - the Lithuanians did not had such a stamp in their permit...)

I got back to lounge and I met a guys from Lithuania who were travelling with permissions and we exchanged with information and showed to each other the documents we have. They were leaving with previous train.

Shortly after they left the Police lady came to me and asked for a permit. I showed just a one part of it - the part with no personal names on it and explained that the second part is in Lhasa with my tourist agent (I lied) and the company who was just there took the rest of documents as the group they're travelling is bigger and They told me that this document will be enough for me. Police lady asked me for a passport and gave back a permit and a passport after checking a Chinese visa in it.

The guys from Lithuania had a copies of the permit. I believe it is possible to fake them as they do not check the numbers. Luckily for me...

The train has been delayed for two hours so it is departing at 18:46 instead of 16:46. I afraid to leave the waiting room as I see it as already passed obstacle and I'm not willing to shine around much as the less noticeable I behave the better even if once 8 got through with my lies and a fake permit. I don't like to lie and I feel actually really bad because of it but I guess that in current situation this is the only way to continue my journey.

I hope my permit will work in a train but at least after the police saw and accepted my story I feel a bit more relaxed and have an inside feeling that I do have a permit even if I don't - that inside feeling makes me more powered in my lies. And it's good. Kind of confidence which I was lacking. A lack of confidence which held me back and because of which I stayed in Tianjin for so long as everybody was scaring me.

With my new Tibetan friend we speak using gesture only as none of us understands each other. So even if I'm speaking I'm doing it in my mother tongue -Latvian. The monk knows that I'm travelling without a permission and it seems like he does not like it, but ... Life is life!

But finally - I got on the train which is very modern comparing to the one's I've driven before. It has got much more space in a doorways and it seems like it's brand new.

My real journey begins. From this moment I'm illegal. I'm a smuggler who smuggles in myself. I'm going to wait until the next station which might be in about 3 to 4 hours and then sneak under the chairs so I could continue my trip unnoticed until Lhasa - the beginning of HARD PART II...

Just noticed - about an hour after my entrance into a train the police patrol containing three policeman where doing a checkoff documents and tickets. I went in a doorway for a cigarette just with my ticket in a hope that when they will get till me I will tell them that the rest is in my bag and hopefully they will skip the detailed check - but it did not worked. The police went back with me in a saloon and checked my passport, visa and ... permit, off course. AND I PASSED!!!!!!!!!!!!!!!!!!!! :D:D:D

Thanks' God! So far - so good.

Again, I have not slept all night even if this time I had a seat and the train was less crowded. I was celebrating the fourth week of my travel with a green tea and I guess I had it to much as I was as an even at 6:30 am I felt just like after the gym.

The train from Lanzhou stooped only in Xining and then Tibet - where half of the people moved towards exit. By my huge surprise one girl was passing me by and welcomed me with a greeting in English: "Welcome to China" and made me a gift - a very beautifully decorated hairbrush and it was soooo unexpected that I even lost my tongue for a while!

Moving further towards Lhasa the area gets mountaineering and it's already possible to see some snowy mountain tops and train administration passed some health check notes. Also it's forbidden to smoke as it might be dangerous for a life - as provided information at the train speakers... The highest point where train gets is about 4700 meters above sea level.

Around this area a lot of antelopes, wolves and eagles can be seen from the train's window by the railway. But generally I got used to the beautiful scenes and after few more hours it and stopped wondering about it...

I found out that the most Chinese doesn't understand Tibetan language and probably that is a reason why my permit was accepted!?

But somehow I got till Lhasa. In the train station was CROWDED with the policeman!!!!!

They were looking around - i just somehow managed sneak till taxi, jumped in and away from there till Lhasa. And it's full of Police as well!!!!!! I cannot stay here as in each hotel they are asking for a permit - I continued moving towards my destination.

It is much harder than I imagined. Much harder. It's raining outside and I'm so f**** scared. I'm even shaking. Found some restaurant with WiFi. I don't know when will be my next post...

So now I'm heading towards mountain Kailash which is about 1200 km from Lhasa West direction. It's a strictly military territory and I need to have some other permit. this fake one I got is only good in Lhasa - but if I will be stopped with it in here - it will be bad as it is not real and here they might now that. It was only good for a train.

My Chinese visa expires in less than 20 days so i need to hurry up! I was planning to leave Lhasa same night. And I guess I Just did not what I'm going for. But let's see! Because after getting to Kailash about 1200 Km from here I will need go back about 800 to Nepal, from where I will be safe and will continue my trip towards India.

Relaxed my nerves while typing - will be easier to walk. I plan to walk around 50 km tonight and start hitchhiking in a morning as the further I get from here - the better I guess. This Tibetan police is scaring me!

Yesterday I started my trip towards Kailash and after passing two checkpoints I seriously stuck on the third and I had to return as it was impossible to pass the third one. Unfortunately this trip was to crazy and to full of action to take any pictures from it except my feet afterwards and a plan of a border I draw. As I'm traveling without permits - I can't leave the territory of Lhasa with the fake one I got.

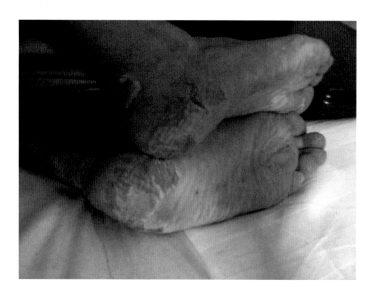

Outside was raining heavily and I was all wet. The first one I passed in a car that stopped by herself seeing me on a road in a night. I just hided in a car so that the police could not see me. On the second there was nobody and afterwards the driver was driving another way and he let me out some 7 km before third checkpoint in some military base. After walking on a road I saw a checkpoint and decided to bypass it through the forest from the right hand side. So I walked around an hour until through some rice fields finally got till some forest road where there was a big stone courtyard from one side of it. It was obviously built for the people like me who wished to by-pass it.

It seemed weird that even if the courtyard was high it was very easy to get on it which I did and noticed the second courtyard coming right after the first one with a gap of around 3 meters in between them and after that a huge field. I thought it was a kind of a trap and I was right. Before I wanted to jump into it I took some tree and checked the ground in a gap - it was just looking as a ground but actually it was something soft which went down at least half meter deeper than it looked like. The second thing I noticed was that the top of the courtyard was unreasonably big - and rounded - so that if someone jumps into it - it might be impossible to get out from it. Thirdly in the field after the second barrier was a projector lights as in a football field which were turned off which gave me the idea of that it is definitely a trap and probably if somehow I would pass the first courtyard - I would face that soft-land trap, if I would pass the second - probably some clever Chinese system would turn on the lights and I would be trapped in that field.

So I walked around half hour by this massive stone construction until I reached the mountain which I think nobody would be able to overcome without a special gear. So I walked back and to the other side. It was already about 5 AM at it was raining all night. On the other side of the courtyard wall was a small border patrol village. When I decided to bypass and begin my move after monitored the life in a village some dog started to bark which I did not noticed before and after jumping into a bushes I noticed guard coming out o have cigarette. After he entered back the one of the houses I decided to give a second try and started moving more slowly so that dog could not hear me. As it was raining everything was noisy anyway... But - I noticed some lights coming definitely from a car and I ran back and i a last second hided behind one of their houses as there was no time for running further. It was a change of guards and a car was driving them from base till border control.

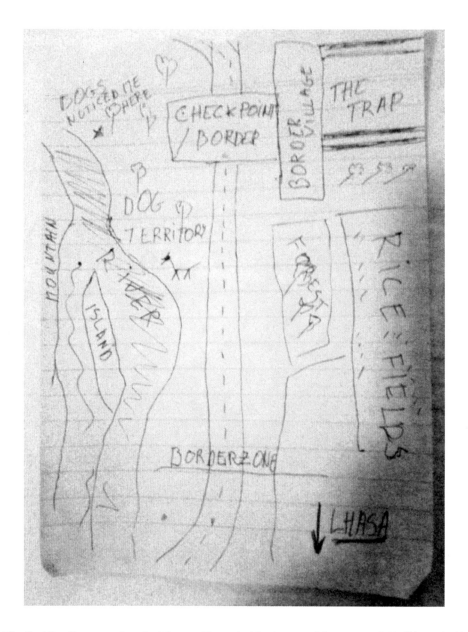

After that the Sun was already rising and I needed to do something very quickly if I wanted to bypass it today. As the life was already waking up in a border patrol village I decided to cross it through the other side as there was a river and I thought I will be able to walk through it.

So very quickly I moved back on the road and crossed it while some lorry just passed me towards border and flashed to the side with lights so they could not see me crossing the road.

I into a forest, got till river and was slowly moving towards the other side of the border right beside a river until the moment I heard a dog barking and I froze immediately and I saw three army dogs running towards my side. I stood frozen and even did not breathe as I know that dogs are unable to see a stood body further than 50 meters. One of the dogs seemed to be a leader. He stopped in about 30 meters before me and looking towards me barked one time then turned away from me and all three of them ran barking on other direction.

Only then I understand the reason of absence of the grass in that forest as everything was ran off by dogs. Also I understood why there are so many shoes, jackets and bags around...

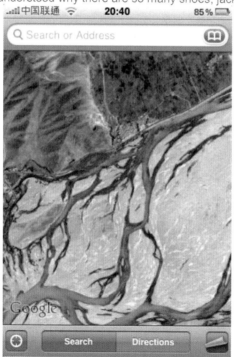

I moved back on a road. But it was already to light to just walk on it unnoticed and I was on a border land so I could not just walk on a road from a middle of nowhere as a guards could see me up to around two kilometers as it was just straight toad with no trees around as they were cut off. So I moved back by the river and was actually half way inside a water with all my bag and clothes. But it was better than rather get noticed by border patrol or get caught by so well trained dogs - which means they were specially trained to control that territory and did not noticed me probably just because of a heavy rain.

So after walking around half hour into a river I finally got out on a road and hitchhiked some tractor. I was all wet I it was cold. Outside maybe some ten degrees. And after I hitchhiked some lorry who dropped me back to Lhasa passing two checkpoints hidden. In Lhasa I went to a travel agent whose contacts gave me the guys from Lithuania I met in a train station towards here and we made a deal with him - I create a website for his business and he makes me all the permits needed as I'm travelling almost without money - at least not enough to pay for permits which is about 400 euros. It's a deal! So I'm staying in Lhasa for until the permits will be made which might take up to 5 days. It's a good deal and I'm staying in some local hostel for EUR 5 per day. And it's cool here. And it's free Wi-Fi in here!!! Luckily my phone was wrapped into a plastic bag...

It's cool in Lhasa - it's a cool place just to live and enjoy life as everything is on incredible chill... Needed some rest - went to downtown foreigner pub and I'm having some beers and playing guitar. Needed some coming back from yesterday's night.

A bit of a difference in every day's traveling...

After waking up and posting news to the website I shaved my head and beard to zero, at 7:30 I left home towards "my new job", walked by the Potala Palace and felt great! Then I went to the flower shop and bought white Lilly's for the Tibetan girl I noticed yesterday in some shop and could not take my eyes from her. She is sooo beautiful...

Afterwards I moved to my new job. I was seated in my own office and started creating the website. It's cool I can say clearly - during his month I've forgotten how to work and sitting by computer was frustrating me a lot! But I managed somehow and pull myself together and made something. Tomorrow - Saturday -will continue.

At around 11 I went to give flowers to the Princess I met yesterday at the shop where she is working together with her mother so she must be a very special princess if I went there in front of her mother. She is kind of special. She is very, very beautiful. Like very very beautiful! So I gave her the flowers and she invited me for a tea but I refused as I don't speak Chinese but her - English. We had just an eye conversation.

And afterwards I got back to the work and found out that I have brand new client waiting for me! in two days two clients - not bad for a "new in the town"! The first client in which office I'm sitting gives me the Tibetan permits and ticket towards Kathmandu, Nepal - package worth around 300 - 400 euros and with my second client we agreed on 150 euros for a website - so during my illegal stay in Lhasa which probably will last until next Wednesday I'll earn around 500 euros! Which is not so bad and I will even have some money for further travelling!

At 4 PM I was running back to hostel where I had a date with a very nice girl from the same hostel I live in - I met her yesterday at the bar and she was so kind that invited me to see some local monasteries around Lhasa where usually tourists don't go.

So at 4 I went with her - we climbed mountains, laughed a lot, she taught me Chinese, we drunk some tea in some monastery , and had a very great time. She is a writer and very nice 21 years old girl. She also doesn't watch TV, reads about six book per month, listens to Bach and Beethoven and also believes in God in her own way unattached to some religion - so many common things we got so I spent a really nice time.

Afterwards - just before the rain was coming down- we reached Lhasa driving by bus and rickshaws and went to the bar to have something to eat and I had a beer as well.

I'm going to sleep earlier today as its working day for me tomorrow and I got up very early today and feel already tired. It was a great day! First months private party in the way I could not imagine better! Summer - the girl who were guiding me towards mountain passes made me a real good time! She was very caring - brought some fruits for me, was always adding a tea in my cup. She is very nice girl!

Today is a one month since I'm travelling. I waked up before 6. Now looking back on it it seems like a crazy trip which in some places was even more than unreal. I've received so many help from so many people from so many parts of the world. There was more than a hundred of people I met during this time - and few of them have become like a close to me - like the Denis from Kaluga, Russia in the very beginning - I spent about 4 days together with and Aleksey from Zabaikalsk, Russia I had a deep talk and only thanks to him I'm now here where I am or Ingus from Tianjin, China - I knew him from Riga, Latvia - but we never had any relationships before.

Also the paradox is - I went out from home with 10 Euro in my pocket - 50 gave me my friend from Brussels Maris - so I left Brussels with 60 Euro, but now there is about 100 in my pocket and I am in Tibet where only permissions cost a capital comparing to what i have.

So this week I have to stay in Lhasa. Today and tomorrow I will make a website - incredible job opportunity in Tibet to get a permits. Today in afternoon I have a date with a girl I met yesterday in a bar - we will climb some holly mountain today in the afternoon.

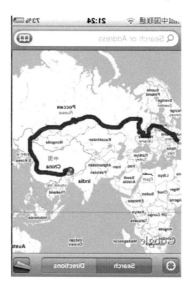

But anyway - I am happy that I am making this trip. I've seen a lot of places and people and have got myself into a lots of different situations which have opened my world more widely. I'm grateful to everybody who helped me on my way and grateful to God who is always watching me and giving me the best situations and people I could ever have.

Woke up at 8:20, needed to go to the work... it sounds funny for me... yesterday's vodka didn't affected my morning at all. so quickly washed myself up - went to the Summers room where she was sleeping in some beautiful dream and gave back a charger from iPhone as I forgot mine in the office. Walking to the work I need to pass Potala palace - I never thought that I will work somewhere in Lhasa and the building which just some time ago was like a Seventh Wonder to me now will be just some house on a road...

Day started on a chill again - came my Boss - we went to the restaurant to celebrate his companies' appearance in the Lonely Planet. Afterwards I got back to the work to continue creating the website. The assistant of my boss is now my assistant as well. He brings me tea every time he enters my office and see empty cup. Also he brings me the food! So at the end of the day I finished the website - www.tibettravelpermit.com - so earned the possibility to continue my trip.

All the necessary permits should be ready in the few days. I cannot go to Kailash anymore as my Chinese visa expires in about 10 days and I cannot make it in such a short period of time. So I will continue to Nepal, Kathmandu which is a few days trip after will finish the second website and will have all the permits. No I'm going to meet Summer and we will go somewhere to relax.

She came down right on time and she seemed to me pretty when walked down the hostel stairs. Once again I found out how beautiful she is inside. She is naturally kind and generally a good girl.

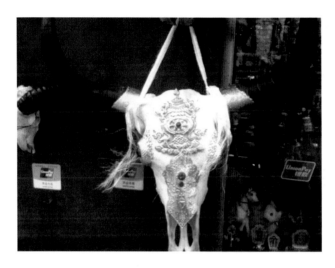

We walked around Lhasa's monasteries and took a pictures everywhere. Just the thing that I really emptied myself out today while creating website as I was making it in a very quick way but it's like creative process and it takes a lot of energy so I felt really tired when we were walking and after two hours wanted to go back home.

In one of the Temples people where bowing down - all of them - starting from small children and ending with old woman.

I did it for a few times as well and actually felt amazing joy doing it and would continue if would be alone. Probably I will go there one day by myself and do it for some longer time.

Afterwards we went to some woman monastery where we were sitting in library where woman-monks where chanting sacred texts. One of the monks gave us a tea made only for monks and offered to the Buddha.

Then we went to some other monastery where we will go tomorrow morning as I have a finally day-off. We plan to spend like whole day together tomorrow - visiting some more monasteries outside of Lhasa.

About Summer: we agreed that from today I quit smoking and she stops eating meet. I have some naughty thoughts coming in my mind when she touches my hands sometimes and comes closer to me. I'm trying to stay-cool. Need to spend more time in monasteries...

Lhasa is a very small city - maybe in 40 minutes it's possible to walk through all of it. The food is cheap. in the places for a locals today both of us with summer had some Chinese food and a lot of chai (Tibetan tea) and we paid only something about one euro. But in a places for tourists one person can eat for about 2 euros.

I like the architecture of the city - houses are small and beautifully decorated. Locals are going around the city and chanting mantras and here are hidden a lot of small monasteries.

On almost every third corner there is some military blocks with soldiers and guns. it's like forbidden to take photos of them as they can just brake a camera. They are well armed and have a guns. Also police is after each 2-3 minutes: very often!

Potala palace is beautiful. To get into it need to buy a ticket one day before as there are limited number per day who can enter it. Museums and tourist attractions are expensive here - the same as in Europe. Entrance in Potala palace costs 10 euros. Daily 2000 people are visiting it so daily Putala palace earns 20,000 euros!!! Guaranteed...

Here is no Dalai Lama anymore - as he ran away to India some time ago because of misunderstanding with a local Chinese government.

The taxi is cheap. Almost anywhere in the boundaries of the city it costs only one euro. The food is cheap here as well. It's just a problem to get something pure vegetarian as everything is so meaty here that even if I take something vegetarian - it has definitely baked or cooked in some animals fats. So I'm staying mostly on noodles and humus.

Maybe I like the city because I got lucky with a guide in it. Summer takes really nice care for me!

I'm going sleep earlier today and tomorrow will try to wake up at six to do some morning prayers together with monks at one of the local monasteries.

I waked up at 7 am and at 7:30 met Summer in the bathroom where both of us where making morning wash-up. It was a cold morning as it rained all night and even when we went out from the hostel it was raining. Summer took the umbrella and we used it on our way to the monasteries where we went to listen to the morning prayers.

Afterwards we went to TaJo Temple the biggest temple in Lhasa after Putala palace and as it was still closed we walked around it and on our way visited a lot of smaller temples. We went early and everything was still closed so we sat down at some Tibetan cafeteria and had some chai - Tibetan tea with milk.

Each monk where chanting morning prayers and in one place they even chanted prayers on us before washing our heads with water. And pouring it in our hands giving to us wash our mouth with sacred water.

We bought some special things which everybody puts in a candle place in temples and monasteries and after visiting some 10 temples we sat down in one Buddhist monastery to listen for Buddhist monks chanting. I fell in some meditative state and time seemed to pass quickly.

Summer was sitting right by my side and from time to time touching hair of my left let - twisting, brushing, pulling and playing with them. I liked it stronger than my meditative state. It gave me more pleasure for sure...

Afterwards we went to some local "kind of McDonalds" Named Dico's to have ice cream and French fries. It was not good. Chinese food is better.

And afterwards we went back to our hostel to take a nap. This was a great spent Sunday's morning!

Afternoon was busy. Me and Summer went out to have some fun. Streets where flooded as the rain was going for over 20 hours non-stop. We eat some just about made crisps and drank Tibetan chai. That was very delicious!

Afterwards we walked in park and later on went to cinema but it was closed already so instead of that we just had fun in local attraction center.

Generally it was very cool day. I spoke with my father and mother on the phone for like about half hour. Could feel their stress about me but I tried to explain that I'm as fine as never before.

Today while walked alone I understood that China is enough for me to want to learn Chinese language. This country has got everything I need accept vegetarian kitchen - which I see as an option for business here.

Tomorrow I'm going probably for a final day in the job - I hope I will finish website tomorrow so Tuesday could go and see something outside of Lhasa if my permit will be ready.

I bought a new wallet for me in a red color with golden stripes for 2 euros as Tibetans believe that in the wallet of red color more money will be. The beginning is not bad!

My day started at 8 when I waked up, took a shower and then went to work. Finished website: www.tibetgtravel.com and earned 1500 Chinese money or about 150 Eur more so now I can continue for sure my trip.

The permits probably will be ready tomorrow. Boss said can leave earliest on Thursday. Feeling tired. Will go now to SPA to heal my legs and... Probably get some massage from Summer as she is coming with me. Well ... What can I say... Nothing much today...

Except that Chinese woman know how to treat the man right! I'm more than a high just because I'm in China.

I'm actually in Tibet - quite sacred place?! ... Somehow I manage to spoil everything in any place ... I guess that is my nature? No more comments regarding this night in my blog... some information should be private..

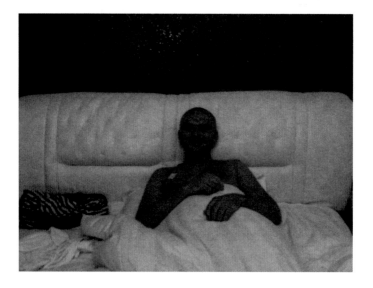

Waked up at 12... feel a bit sick. I have got some kind of flue. My nose is leaking and I have a cough.

Went till Summer and we both went together to one of the monasteries in the mountains out of Lhasa where is only one monk in the whole monastery.

We talked a lot with him and drunk tea. He made some prayers on my head and gave me as a gift beads and said to chant "Om Mani padme Hum".

Afterwards went to post office and sent some pictures to my parents by post. It will take 2 weeks for them to receive my letter!!!

Now I will sleep for about hour and then will go to work probably to get my permits and it means soon might leave Lhasa!

I waked up around ten because of some neighbor who played drums since the very morning. I'm sick and I slept under two covers and inside a sleeping bag.

After waked up - went till Summer and we both went to the monastery up a hill to give to the monk photos he and we took yesterday about his monastery.

We spent more than two hours in his temple and afterwards moved more up the mountain to one another temple just about hour walk up a hill.

The temple was very poor as it is so high up the mountain than very rare who goes up there. But they gave to us everything they have even cookies offered to Buddha so we can have his energy.

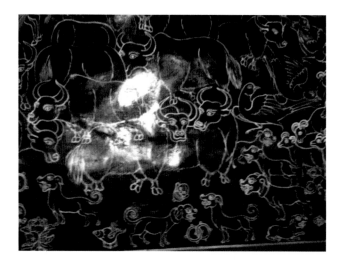

They gave a beads to Summer and told to chant Om Mani Padme Hum mantra all time at least once a day as it removes all sins.

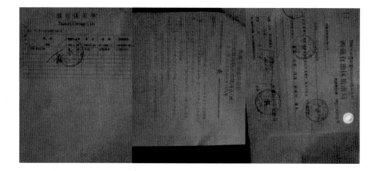

On my way back I went to the office where finally my permits are ready. Still I cannot hitchhike even if one of the permits I got is military. So my boss tried to manage for me a car and if everything works out good I can finally leave Lhasa on Friday.

However I'm leaving. I got the permits but still the transport is to expensive for me so I bought a bicycle for 600 yuan and will go with it. I have about 700 km to go till Nepal border in 5 days. Will do my best. Also today I went to the princess in the shop I gave flowers few days ago. She was there with her mother and her sister, real beauty, too! It seemed like they are twins! soooo beautiful! I will have to write her a letter in Chinese. Kind of "love letter"... maybe tomorrow...

So Yesterday I pushbiked away from Lhasa on my new bicycle and now I have reached Shigatse - in the middle of Tibet. The road was hard and lonely as I was driving through all night. Some youngsters on my way throw stones on me and I realized how hard it was to push bike when I had to drive faster than they can run.

I have not slept all night and not have the power to write much. In general that what I'm doing now is also illegal even if I have permits - as accordingly to the law in this territory I have to have a guide all the time with me.

Just that half road was a real off-road. I illegally passed by 9! Checkpoints! And now staying in some cheap hotel. I can't feel my legs. I have not got off the bike for about 20 hours! And made about 270 km. trying to look more like Chinese to pass borders!

So, after yesterday's biking till from Lhasa to Shigatse around 270 km and staying overnight in a hotel I feel better but my knees are still in pain. I decided to stay one more night in Shigatse and take a place in a jeep to drop me until the border of Nepal as people are saying that the control closer to the Nepal border is more serious and that it won't be so easy to bypass them as I did it this time.

Secondly my legs don't feel so good to drive twice as much as I've driven until now in three days - as my Tibet permit expires. Yesterday's biking tour have touched me deeply. I never saw before such amazing views. Himalayan mountains are so beautiful. From one side Mountains, fast river Yandze, then yellow crop fields surrounded by pink flowers - amazing beauty!

The air is rarefied. That made my biking tour harder than I was expecting. Sometimes I just stopped and was breathing heavily for few minutes. Sometimes I just could not move any further by bike and I walked for 10-15 minutes and then continued pushing the pedals. Checkpoints - that was a funny thing: first was right behind the Lhasa. They checked every car. And I was just driving in the middle of the road and bypassed everybody! There was a lot of! Maybe some 30 policeman...

Then from Lhasa until Quxu - where I stocked in last time - in between there was three more checkpoints which I did easily - just drew straight thru them without stopping or looking back. I think that bicycle is the best way to pass them - as it seems impossible that some foreign tourist would drive a bicycle in a rain in Tibet in the middle of nowhere.

After that started mountaineer are where it was very hard to go up the mountains. But ot was fun going down! I passed a lot of small country towns where dogs where barking and running on a road to bite me in legs. And after towns - so beautiful mountains!!! Amazing views!

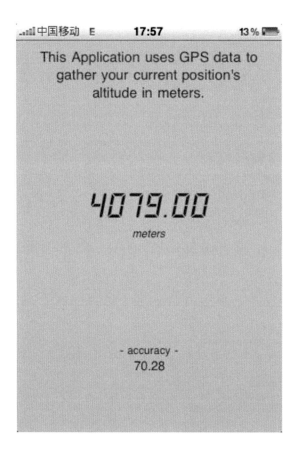

The next checkpoint was only after around 50 km and it was closed one. The police saw me but did not said nothing - they just opened a barrier for me. And after that the bad road started. A lot of small towns and military camps. Everybody told me hello and I was wondering how do they know that I'm a tourist if I was masking and I passed so many checkpoints unnoticed but these people recognized me at once.

And in a late evening I arrived in Shigatse. This morning I waked up from somebody opened my doors and then closed them - I guess - housekeepers.

Afterwards I woke up, took a shower, washed my clothes and went to a local monastery - Shigatse Tashilshunpo monastery where the entrance fee for foreigners is 55 Yuans or about 5 EUR. Very beautiful and big with restrictions to make a photos and some forbidden areas.

Afterwards I contacted my tour guide and we decided to continue this journey in a legal way - tomorrow noon I'm taking a jeep to Nepal. Today - i will enjoy this city! And heal myself somehow...

I found out that I do like Tibetan music and art!!! I spent afternoon going to some temple which seemed like miniature of Potala palace in Lhasa. Later on I found out that this castle was founded by first Buddha to be something like a place where keep real Buddhism and with time Potala palace leave as a museum for Buddhism. It was about 6-7 centuries ago! Already by that time they new about Lhasa to become a place of tourist attraction! Unfortunately I did not saw any monks around this magnificent building. It was closed and guarded by one army guy.

Later on I continued my journey up the mountain behind this Shigatse Fort. I had a slippers so even if the mountain was not for professional mountain climbers - for me it was an action. I felt twice: once upwards and once on my way back down. Both times nothing special - just a little bit more blood on my already bloody legs.

Also I had to buy an oxygen balloon as my altitude sickness got worse and worse. Especially after 20 hour biking from 3200 to 3700 m without previous exercising I feel quite not so good...

At the altitude above 4000 m I meditated. I made a small hill from stones where each stone was assigned to some people I know. Parents, relatives, close friends, people I love and loved, people I know and people who helped me on my way till here where and people I met on my way.

Afterwards I chanted some prayers on behalf of them. It was good. I united in my thoughts with them and united myself with the Highest. That is a living worth feeling! I think a few have got a prayer for him chanted at the altitude of 4000 meters in Himalaya Mountains in front of building founded by first Buddha! Afterwards I got back down to the city and had some food.

Nobody speaks English in this city. So I'm speaking Latvian - my mother tong and they Chinese. Somehow I manage. Not so bad! the biggest problem is to get a vegetarian dish - as almost everything Chinese and Tibetan are making contains rather eggs ether some small and unnoticeable peaces of chicken or pork... In Tibet popular is a meat of a yak - local mountain cow. So sometimes must be cautious about even baked potatoes can be baked in a oil of yak meat.

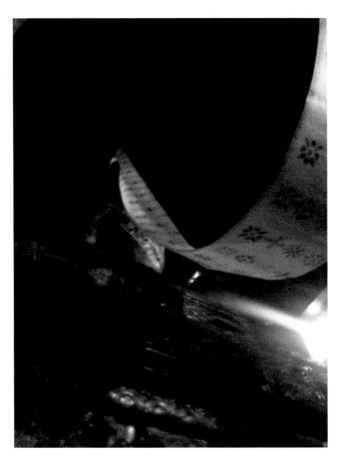

Tomorrow about noon the jeep is coming from Lhasa to pick me up towards Nepal. This joy will cost me about 100 EUR. So I'm happy - even 100 EUR is at least 5 times cheaper than any offer of visiting Tibet. I would like to hitchhike - but it's too dangerous for my Chinese visa.

So tomorrow at noon I'm leaving towards Kathmandu, Nepal. Finally!!! I've spent already almost one month in China! That's too long!!!

Sometimes I feel very lonely on my trip. Like today. Yesterday I received a letter from my business partner Vadim I have not got in touch since Brussels when he visited me about 2-3 months ago because off afterwards he went to India and just now he got back. So after reading his letter I felt touched for rest of the evening so much I could not sleep for half night.

Victor - my close friend - recently he wrote me a letter, too, where he said he's very happy for me and that he has understood the meaning of old Russian proverb: Better have hundreds of friends than hundreds of rubles. And that's so true!

Vedas are explaining that everything we give away from pure heart towards other people is coming back to us when we experience the bad times. During this trip I felt so much the meaning of donations: each time I received some help From people I could remember the situation when I was giving away the same help to other people a while ago. So I understood the deepest meaning of donation. It is like a law of donation. And it works perfectly. And no modern and ever-changing science will get to the point of development to be able to explain it. As it will never explain intuition, materialization of things, reiki healing and e.t.c. ..

So I'm giving even more. Each time I visit some temple I'm leaving as much as I can afford even if I'm on low budget. Because - actually - everything we give away - comes back to us. Everything.

So it's better for us to give away the best things we can give: Love, Patience, Forgiveness, Gratitude, Hope. It doesn't cost much - but these are more personal and meaningful donation when turned towards people than just a money left in a temple. But sometimes they cost us much more than we can imagine: to let go Ourselves - our understanding of who we are and how we are. Sometimes donation means donate some of our qualities we used to think is ourselves. Donate and let go of our Ego is the greatest donation everything rest will come from. Our Ego is like a stone towards a way of a love powered seed in the land willing to rise above the ground, spread the leafs to produce oxygen for all living beings, blows for those who understand their beauty and nectar - a gratitude to insects who spreads the seeds of the plant further.

We where borne to be good. We just need to remove the stone that covers our goodness. Our true Ego is to help others and to do as much as we can for others. Our fake Ego - to do everything we can for ourselves. Funny how sometimes plants are more filled with understanding of their true selves than humans. Now I'm waiting for a car who will pick me up towards Nepal border from where I will drive a bicycle again.

I got lucky again! my company towards Kathmandu in a jeep is a French couple who have traveled all around the world - Leo and Margo. They have a website, too: www.sur-la-route.net and it is interesting to speak on our road about The World! I attached my bicycle to the jeep using one lock-chain, one rubber strap and 40 meters of rope I bought in Brussels for cases of emergency...

Himalayan mountains are out of describable beauty! How colors flows with everything and makes such a harmony together. It seems like everything is in its places - even clouds.

I got in the car in altitude of around 3800 m and the highest peak we reached was when we passed mountain Gyatsola pass 5200 m beyond see level. That's the highest place I have ever been. We took off our T-Shirts and make a pictures in the highest point. It was cold and snowy.

Afterwards we stopped at Tingri at hotel Snow Land, 300 km from Shigatse. On our way there was three checkpoints - but now I didn't had to worry about them as finally I was officially driving with tour guides.

Two of the checkpoints where empty - without any guards in them but the third was the serious one - more serious than any other I saw on my way.

It was not just a regular checkpoint - it was a military one. All of us had to go inside of it and give our passports where they checked our visas and compared photo in passport with our faces. The guide hanged in all permits and soldiers made some remarks on them and in their journal. On our way we also passed road which led to mountain Everest base camp.

Snow land hotel is a typical Tibetan style building complex with small but cozy rooms for which I paid only 3 euros and a Tibetan restaurant where I had a fried vegetables with a ginger soup.

We were laughing about my similarity with Leo: both of us have got iPhones, both of us has got websites, both of us works as a web designers, both of us have a car charger that doesn't work and we are very similar on our faces - we have the same haircut and beard, and both of us have the same rounded face and green eyes. In my passport photo he looks the same and our high is the same - 177! So many similarities!!!

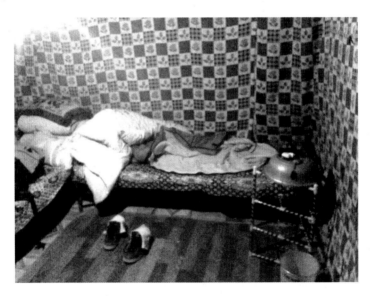

I'm at altitude 4700 m now and to avoid undeserved headache of altitude sickness. Electricity here is bad. It is not as strong as in the cities and it takes a TIME to charge it up. So nothing to do - day 39-40 I'm celebrating by charging my phone and drinking beer mixed with coca.

Probably because of high altitude I could not sleep all night. Finished reading a book about all Dalai Lamas and started one all Dalai Lamas have studied - Tantric one, in my "yak's room"... It's not even so cold! Almost all Dalai Lamas have studied sutras, tantra, astrology and Sanskrit - which is basis of Vedic texts. But "veda" from Sanskrit translates as "knowledge" or "science". It's no wonder that the Dalai Lama we know today have written over 85 books and have succeed in so many ways that his name is known worldwide. It's so important to study the essence - the true science. The real knowledge!!!

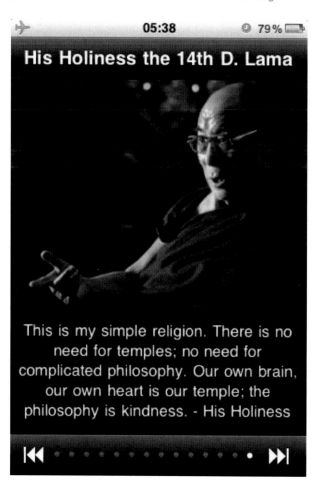

...waiting for tomorrow to see gigantic Everest..

Well - it's far ... I mean really, really far... Maybe 80 km away from where we are, but I can still see his Greatly Glorious Magnificent Massive White Peak Undisturbed by anything Peacefully Drilling into the sky with Ice Cold Simplicity a parting clouds from both sides. The Everest let me knew in the morning who is The King of this area. Just by one look towards it makes me willing to bow before His Majesty.

Outside is really cold! Maybe just a few degrees above zero. I have not slept all night - was sending e-mails to my friends, reading book and reviewed photos. Probably because of altitude. It's about 4700 in here - 1 km higher than in Shigatse where I stayed for two nights and about 1,5 km higher than Lhasa.

At the breakfast I had the same ginger soup with vegetable rice. I know that it's bad to eat rice in the morning but I felt hungry and it's delicious in here especially when the food is accompanied with the beautiful Tibetan music! Meanwhile phone was on the super-slow charger to continue charge up what missed yesterday.

Also I tried Tibetan national dish which Dalai Lama eats every day and recommends to everybody eat it every day against cancer, called "tsampa" made from barley and dried yaks cheese turned into powder, mixed with sugar and added hot water.

Also we made a "Tibetan shower" - on coal made from yaks shit burned juniper incense after which I felt really great - my nose opened and I felt fresh energy coming inside me.

Actually it's amazing in how many ways Tibetans are using yak -they make even the fire from their shit dried in sun instead of wood! They collect shit all summer, dry it and in winter burning it.

If yesterday I was dreaming of leaving this tourism business obsessed country then today I feel sad each minute going more and more out of it. I experienced a part of the real Tibet. It is something different than tourist filled Lhasa. The couple who managed this hotel and restaurant was shining with peacefulness and love. It was a pleasure just to be accompanied by them. The atmosphere was blessed and their business filled with love everyone can enjoy. Great man and devoted wife. Om Shanti Om.

As we are getting closer to the Nepal border the altitude gets lower. Seems like Himalayan Mountains just don't want to stop showing their beauty. Towards Nepal they become more greener and full of waterfalls. It seems like mountains are milking clouds and they happily gives away the best water that ever possible to get et on earth.

Mountain serpentine are crazy. Very often we drive under some waterfall, bit sometimes there are tunnels when waterfall is to strong. Driver knows roads well, but anyway even with a speed of only 50 km/h, looking just behind the shoulder and seeing nothing down for at least 300-400 meters except rear clouds makes me feel how much currently my life depends on drivers mastery.

Just before some 15 km before the border we saw car crashed and our guide told us that two days ago there was a big bus who felt dawn the cliff...So after two more kilometers the border zone started and I had to take of my bicycle from the back of the car and drive down the hill by myself. It was the best part I ever experienced driving a bicycle! 15 kilometers down the hill!!!

I made a lot of pictures and even managed to make some videos! at the Nepal border I waited for them to come for about half hour and after we met we went to some restaurant for the last time have Chinese noodles in china.

And afterwards we crossed the border. It was full of people caring some heavy stuff on their heads, shoulders and - basically anywhere on the body was possible to put something on - there was something...

The difference between China and Nepal is HUGE!!! First of all people - since the very border we have been followed by two Nepalese people who wanted to give us some help even if we refused for 20 times.

From one of them we got lucky to get rid of after I turned to him straight walked eye-to-eye in a distance less than 5 centimeters looked straight into his eyes and told him that I don't want to see him anymore following me and got sure he understood it.

To get rid-off the second one was harder - he had some connections on a border and when we stood in a lane to get our Nepal visas, which costed about 25 EUR for each, one of the border controller came to us and said that we HAVE to use this guys offer if we want to go to Kathmandu. Quickly we figured out to say that we are not going today but tomorrow as we had a long day and it's enough for us to travel further today and we will stay in some local hotel on a border.

So after each of us got Nepal visa, we moved forwards and through hotel found a jeep who could bring us from border to Kathmandu: three people plus bicycle for 50 EUR, which for us was OK and we attached bicycle on the roof of the car and comfortably begin our move through the incredible fairy-tale look-like country side beauty of Nepal until we reached road block caused by land-slide.

We stocked on a road for about hour. I took a quick "shower" in a nearby waterfall and felt good! Meantime local children managed to gather around our car and check everything what was in a back, what was on a roof and even managed to ask us a money which we had to refuse.

The landscapes are beautiful in Nepal. Everywhere waterfalls, hugged in between green and huge mountain valleys, surrounded by different kind of trees and plants and gigantic butterflies!

There was one car which could not pass the valley after road was partly rebuilt just to continue the cars moving. Everybody was helping them up towards the hill. We came, too. And I saw 4 people sitting in a car. I was like wtf!? And we opened a doors and said them get out. And with a big smile they got out and finally there was enough power to move the car up the hill.

And afterwards we moved forward to Kathmandu, where we arrived in the very evening and moved in our hostel. Instead of paying for transportation I left my bicycle to the driver as I thought it is a good deal - I lost only about 15 eur, but had an unique biking experience in Tibet. It was like a "rent a bike" option. So now I'm free again! We were tired and after supper went sleep.

I waked up early morning to go to Chinese embassy to give away passport for my second Chinese visa as this one I had was not enough to go to Mt. Kailash for what all this trip was generally designed for. So this time I applied for two month visa in hope that it will give me enough time to go to Mt. Kailash.

Afterwards I went to Monkey Temple - or Swayambhunath in Indian language. Locals believe that the monkeys living in a temple are saint.

According to ancient Swayambhu Purana, all this place was once filled with an water, out of which grew a lotus. The valley came to be known as Swayambhu, meaning "Self-Created." I finally saw the "eyes" I've been willing to see it for a last three years!

I like Kathmandu - everybody is on a huge chill here and I'm happy I got rid in a good way from bicycle! it's so much easier to travel without it. I'll stay in Kathmandu until Friday/Saturday and will move on as soon as I'll have my passport back from Chinese embassy with my Chinese visa.

Today I bought two rings for me: one with a pearl and one with smoky topaz. They are beautiful!

I did not went to any place today as just wanted to chill and relax. So nothing much today..

Today is the 1st. September - and according to the moon phase - birthday of Lord Krishna.

So there is a big party and we are going to one temple later on with some girls from Kathmandu - local guides.

We visited Patan temple territory and stayed there for a while going around them and inside out.

Everything was very crowded: woman wearing red saris and men as happened but in the air was full with celebrating mood.

Today was raining whole day so I was sleeping half of it. In the morning continued reading the book about Kundalini, probably will finish it tonight.

I went out for a pizza and using their wi-fi through my iPhone called to my mother and father, sister and few friends and afterwards went back to sleep in Hotel and prepare for tomorrow.

Next day early morning I went to the Chinese Embassy to get my passport back. I'm happy - I got the Chinese visa for 60 days. That is good!

After that was raining all day so I cancelled my trip towards India until tomorrow.

Later afternoon I went to some meetings with local businessman to talk about web development and after went to post office to send a parcel to my friend Victor as he helped me a little bit today.

So nothing special today and I feel prepared for India. Tomorrow - Step Three in my travel diary starts - India!

Morning started as usual - waked up, took a shower. Packed my bag and said bye-bye to my new French friends Leo and Margo. As I am totally out of any cash - they gave me some money to continue my trip. I left Kathmandu hostel Holly Lodge at around 11 AM.

I was going out of Kathmandu for few miles and after I took a local bus until the next city. Then I continued to walk through beautifully Nepalese part of Himalayan Mountains until I hitchhiked one truck who was so kind to drop me around 150 km through mountain serpentine.

The Himalayan mountains are very green and full of life at this part. We even stopped at some Nepalese road cafe to have some tea with milk.

As it was already dark in Hetauda truck driver seated me on the bus going to the Nepalese - Indian border-city Birgunj for only 80 Nepalese rupees which is less than one euro for more than 50 km. He explained me that it is not safe to go there so it's better that I take the bus.

The "bus" was a jeep in which somehow was sitting 12 people including me. We were driving really fast and in about hour reached the border city.

In Birgunj I arrived at about 8:40 PM. The border is closed during the nights and it will be open only next morning 9AM - until

Then I had to find some place to sleep overnight. And I found some hotel on a road where instead of asking price 450 rupees I got the room for 250 which is about two euros.

The room reminds me more a prison cell with two separate beds inside and additional room with toilet and shower. No covers for beds and it is definitely that they do not change they bed kit for a lots of clients as everything is in spots and stinks. But it's okay for me as I have got a sleeping bag and at least here I'm safe.

I waked up at 7:30 on my bed moved away from the wall so to minimize the possibility for creatures on the walls to get until me. Took a shower and went out.

The city was alive already by that time. A lots of rickshaw offered me a ride until the border but I decided to walk my last hour while in Nepal. Had some Indian breakfast on my way and very soon I was on the border.

The Nepalese side wanted to charge me 300 rupees for service charge - I gave only 25. Then Indian side charged me 300 Indian rupees which is about 5 EUR I guess. So further I'm hitchhiking!

I'm in India and heading towards Varanasi. Cars picked me up very easily but the driving speed was poor and distance usually about 10 kilometers. so driving forward takes time. Also the water on my way was not a problem as all the small villages had a pump so I was filling my bottles in each of them. It is enormous hot. I'm sweating so bad that I'm all wet.

I got lucky with one truck which dropped me about 150 km. We even stopped for a lunch brake and driver - nice young man - paid my lunch. After the lunch started a new part for my trip.during the lunch as we talked where I'm going and driver said he soon will go sleep as it was already dark - one of the guys from road cafeteria offered to help me out by dropping me until Varanasi for 500 rupees. I explained that I don't have money at all but it seemed like they just did not or could not understand it or believe in it.

After trying for about half hour to explain my situation and absence of problem in it he guy finally told me to help for free. He was a local. After the lunch he invited me to his home and introduced me to his family. They where asking a lot of questions and it was clear that they do not hitchhike!

Family was friendly but I was scared of next day that they just could not accept hitchhiking as a driving method and wanted me next day to drop me not until the next city but until traffic police. But generally family was friendly. They offered me a night sleep which I accepted but ran away after heard everyone dreaming. So I continued to hit a road at about 3AM and soon hitchhiked my next car.

It was some rich local oligarch driving in a back of a jeep with his wife. I was seated in front with a driver. Then I had to explain the reason of my travelling and why I travel with no bus or train. But it seemed like it is out of possible understanding of Indian people - hitchhike; no money ... No train ... It seemed weird to them. So weird that they stopped twice at the police checkpoints and Oligarch was explaining them each time the obstacles of his worries - me. They dropped me about 100 km and the road took about one and half hour. And during this time I was unsuccessful in explanation that I'm travelling without money even if I told it 10 times and showed my wallet, and told that it is impossible for me to travel by train - they y dropped me of at the rain station and even asked for a money for a trip. I had to say that I'm taking a train as otherwise they would leave me at the police station. Crazy couple. Incredible misunderstanding... I was so pissed because of them and happy the same time as I had reached Patna.

It was 5 AM. And I continued to walk. Through all Patna, managing to find my way towards Varanasi. Whenever I asked somebody for a way - they tried to divert me towards train station and explained that it is not a walking distance. At the end I was so tired of explaining that I'm hitchhiking that I just used my GPS to find a way. Indians seemed to me incredible stubborn and unable to accept some new ideas. So I continued walking with no night sleep and drunk water from the local water points. I hope I will be alright!

It is very hot. I mean very, very hot. I'm drinking about one liter in two hours and I'm not going to toilet - I'm sweating so badly! The "hot" is not the right word. It is extra super-hot! I felt my bag is to heavy as so many unnecessary things for such a weather conditions and I thought about not only throwing out few things but throwing out bag itself with all sleeping bag as I'm sure I will not use it for the rest of my travel and it is only giving additional weight and so difficulty for my travel.

I stopped at many temples on my way but all of them where not a walk-in but just pass-by types. So I just passed-by.. Even without picturing them.

I continued to walk and finally one car picked me up where was sitting two young people and for my happiness understood that I'm travelling with no money - gave me a bottle of coke and after let me out at some smaller village and paid for a bus to drop me forward some 100 km more. That was a good thing. I was travelling on a roof of a jeep with some other 8 Indians.

After I was walking for about 10 km until finally was. Picked up by some tractor driver and was sitting at the back in the sand box and after he dropped me for about 10 km - I was picked up by some guy who was driving a motorbike.

It was not just a guy! He was a local leader of India Communist party and was even elected as a local governor.

He is a leader of the city. Very warm heart - knows everybody on the street. While driving at the back of his motorbike - I was seeing how he dealt with locals on it's way. He introduced me to a local clergy. Some advocates and judges from the court. And some local businessman. He wanted me to stay in his village for a night and to show me how Indian people lives.

We stayed half day in his sisters house - I was treated like a king. One woman was making a wind for me, another food, third tea, and fourth washed, dried and even ironed my dirty clothes while I was eating and sleeping for about an one and a half hours.

Later afternoon we went to his house where he introduced me to his uncle and his family and after evening chat he pulled out vodka with words: "I know you Russians drink vodka!" i agreed on that at the bottle was emptied in the very friendly company of four people his relatives.

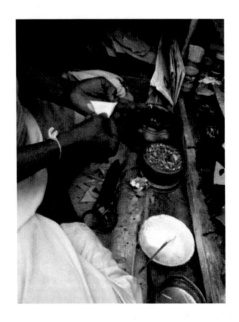

After they made me a bed outside covered with a mosquito net. But I slept very bad because of enormous hot weather I need to get used to. I'm happy for my new Indian friends! very kind and nice persons. Probably I'll stay one more day with them. Mosquitoes are bad in here!!!

The house I stayed for the night is a typical Indian village style home beautifully decorated outside with columns of marble. They continue treat me like I was a king. It is part of their culture to give guest the best they have. The food is excellent. Hosts are taking care of all my needs. Except the wi-fi I have everything! I could stay here for a week's! I just need to contact somehow my French friends Leo and Margo as we decided to catch up in Varanasi. But as I don't have an e-mail access - don't know how much time I have got to get untill Varanasi.

We draw back to Arrha and visited the family of my new friend. I took a small nap, we ate delicious food prepared by his wife and after we went to train station and my new friend - Avimash - bought me a ticket till Varanasi. Everybody is very welcoming and friendly. I found out that behind the annoying shouting "hello" on the streets deeper Indian culture is very kind and delicate towards guests. I'm happy I met Avimash not only because he helped me a lot but mostly because it is good to know good people all over the world. I believe I gain a new friend in India.

After spending around three hours sleeping in the train to Varanasi I waked up just before the bridge upon sacred river Ganga and after 15 minutes train stopped at Varanasi. The closer I got to the city the bigger mess was outside. Everything was stinking even in the train.

I get of at Varanasi and slowly walked until ISCON center - Hare Krishna ashram where I planned to stay at Varanasi for free. And after one and half hour walking through crowded city with constant refuse to rickshaw who were asking all the time to help me finding cheap hotels and dropping me there I finally reached ISCON.

It was like an oasis in the desert. Only instead of water I found peace here. Loss of crowded city noises in peacefulness with Hare Krishna servants. And what is more important - they allowed me to stay for five days. This is my maximum stay in Varanasi in hope this time will be enough to see the temples of one of the oldest cities of India, swim in river Gang, meet Leo and Margo and also after two days - my new friend from Arrha - Avimash, who told me he will come here to show me the city in two days.

Today is Tuesday. Avimash will come on Thursday. I need wi-fi internet access which I could not find in whole city to contact Leo and Margo and find out when are they coming to Varanasi. The commander of the Temple gave me a separate room. I laid in a bad after long trip so could not feel my stomach pain which lasts now for a whole day because of drinking water I used to drink on my way from road water pumps.

In these four days in India I understood that it will be very hard to travel through it without money. As even water suitable for drinking is possible get only for a money. But qt least for next five days I have a place to sleep and delicious Krishna kitchen food to eat. That is very good.

Hare Krishna!

Today is the 50th day since I'm travelling. It feels like a lifestyle now more than a travelling. But sometimes I feel good when thinking of that this will end very soon. Looking back now in a map and explaining to the people what I made to get here seems pointless as I even myself can not believe in it. Now looking back - it feels like I'm lying when telling that I came here where I'm now from Brussels all the way hitchhiking except for China where I had to take a train but together with that I got a lift of about 150 km in China and about 300 km I was driving a bicycle in Tibet. Generally it all even for me looks incredible and unbelievable sometimes. It's good that I have this website - thing to remember and to realize what have I left behind in these 50 days.

I got lucky about internet - yesterday I found it for free and the speed is quite good! My morning started at 3:30 - and I guess I was the last one who got up as everybody gets up at around 3 AM!!! And are preparing for morning prayers dedicated to Hare Krishna starting at 4:30.

I must note that it is my first time I'm staying in ashram and everything is new for me. Even if I don't understand the meaning of a lot of things and probably Hare Krishna as religion is not quite suitable for me, I decided during my stay, do everything what local monks are doing to pray the God.

Some call Him Allah, some Jesus, some Krishna, some Buddha ... - anyway my believe is that God is everywhere and there are just different ways to approach Him. He is everything and everywhere so he is just too big to fit inside just one religion. Every prayer to the God for me is good and doesn't matter the language or the name of Him. So I pray God and was singing praises to Him according to Hindu believe system - so today I was singing praises to God Hare Krishna and it feels great! At around 5:30 AM I went back sleep and be a food for mosquitoes... and waked up only around 11.

So I hit the city and met some German guys travelling like me all across the world for months. We visited a lot of temples and for me most exciting was Golden temple where security was very high and everybody had to leave everything outside. So no photos from that amazing place where yogis are chanting Vedic mantras and the place just feels like shining from blessed and pure energy. Also I was swimming in Ganga. Even twice. Once alone and second time with the local cows.

The guys I met are staying in the Ganapati Guest House. I have to say - it is amazing. Nice Ganga views with rooms windows leading to the Ganga- it is amazing. I recommend it for everybody who wants to go to Varanasi. In such a jot weather the Ganga winds makes air cooler and Varanasi seems like the best city in the world.

Generally - this city is crazy. It is very old. I don't know how old it is - but it seems like thousands of years. You can see temples built on ruins of older temples. Temples everywhere. Small streets. Low doors. Cows. Goats. Sick dogs. Hashish dealers who try to sell everything they got even shouting loud in front of police.

Yogis and lot of weird stuff.. We went to some yoga classes. Generally - well spent day! I'm happy for swimming in Ganga and visiting monasteries. Varanasi is incredible city!

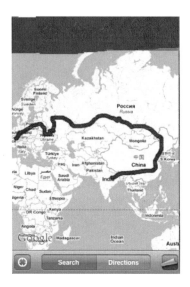

Today I waked up at 8 after well spent night as a feeding body for hungry mosquitoes because of which I cannot say I slept very well. And also during the night I braked the screen of the iPhone by fighting with mosquitoes. It just slipped out of my hands and felt heavily on the stone floor - but it works! So I wrapped it in transparent tape. I think this will work fine until I'm done travelling.

During nights the scent of freshly baked dead human bodies who are later on floating in the same river I was swimming yesterday is coming everywhere in my room as the ashram is located very close to the Ganga.

I was not eating anything yesterday as when I got up breakfast at the temple was already finished and as I did not had the money and was spending all day outside of the temple - I had a food only at 8 PM - when I returned back. But I did not felt hunger. The city excited me so much! But today I went to the breakfast. It was at 9.

Yesterday after I got back to the ashram I got pressurized by local Hare Krishna book store manager who without empathy was reading some verses from Bhagavad Gita and tried me to explain something I could not understand. Even after I went to my room and took a book to read - he found me, took a book from my hands and was reading it like he had red it before loudly to me with his comments. Probably he had red that book before as it was a Vedic book and I believe he knows a lot but I just needed my space to relax from all day noises and invitations to buy some stuff and take a rickshaws so even if I tried to listen what he is saying he was pissing me off badly and it was impossible to be alone. So I finished reading it only after midnight. Great book! Called "Tirikural" - The Great Book of Tiru-Valluvara. And today I started reading the book of Swami Ranganayhananda "Practical Vedanta and the Sciences of Values".

I spent morning in the ashram watching who is doing what and observing temple traditions as I was waiting for my new-found friend Avimash coming from Arha to Varanasi 200 km to show me the city. So I was waiting for him to arrive in the temple.

Breakfast was great. I stop counting how many dishes we had at 8 I guess. There where rice, rice with milk, potatoes with something, dhall, beans, banana and apple mix, chapatti and something else. To be honest - I like Hare Krishna temples and ashrams because of the food they make: it is always pure vegetarian, prepared in clean dishes where never ever since they have been produced was any kind of meat inside. So the food is clean physically. Also those who are preparing food before doing that always washes themselves, prays to God and offers the prepared food to God - so the food is called Prassad - meaning - offered food to God. The idea of prasad lies at Hindu Scriptures Bhagavad Gita where it is said that person who cooks only for himself without offering the food to one who have provided it is merely a thief.

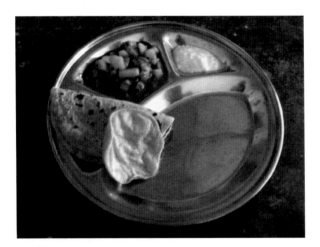

Interesting thing about Hindu believe system is that it is not based on believes. It is based on scriptures called Vedas which has been written during a long time period by sages who have spent all life to realize the higher Truth - the rules of Universe and God. For Hindus the highest realization of Truth is to find that God is inside ourselves, that we are the God. And so these scriptures are not to believe in them but just as a guidance where to move in the process of self-realization.

From here the vedatrac comes from. Veda Track was already taken domain name so I stopped at vedatrac.com and also I have bought a vedatour.com but for this experience travel seemed to be better word than tour... Veda - generally - means Knowledge. Knowledge gathered together by yogis, sages, rishis in realization of highest truth and written down in a manuscripts which ancient sage Vyasa gathered together and made four parts of it - rigveda, ayurveda, samaveda and arthaveda each of them making different realization for better life by explaining laws of life.

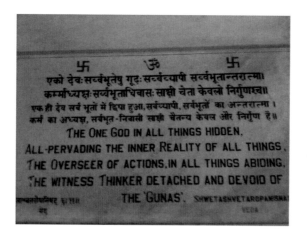

At around 11 AM Avinash arrived at the ashram and we took a rikshaw to drive around Varanasi for a small excursion. We went to Durga Temple, Sri Ram Temple decorated with Ramayana whole story from inside, Hanuman Temple - Manas temple or so called "Monkey Temple" with a security standards I had to leave my mobile in a locker outside and new Vishvanath Temple.

After we moved to Ganga and took a swim in there and had some laasi (yogurt drink mixed with fruit juice).

On our way we discussed some business. It was a good meeting after which both of us moved to their homes - he drove back with train to Arrha and I went back to Ashram.

Morning started at wake-up call at 6. Laying in the tent made on my bed but still I found 6 mosquitoes whose hunger for my blood somehow made them manage to get inside a tent.

Afterwards I killed half of them I quickly took a shower and went to Brahma Kumaris center for morning meditation or to see whatever their offer is. The guy who was introducing all that yesterday inspired me a lot. He was a doctor himself and left his patients for 15 minutes just to briefly explain me what for is Brahma Kumaris - it is like an international enlightenment center doing everything for free.

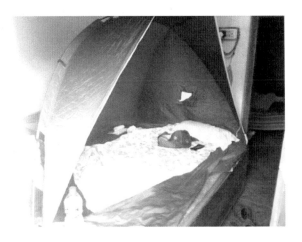

Like written in the Vedas - true knowledge always is for free. I wanted to check it out - but it was in Hindu and as nobody came to meditation except me - it was not happening. So I went back to Ashram and took a sleep for one more hour and waked up at 9 for a breakfast.

After breakfast I went for a walk across a temples. On my way at money exchange point I exchanged my last One Dollar bill and get something around 40 Indian rupees which was enough for me to by a soup and local sim card. That's actually all my needs for next few weeks in India as my next stop is going to be some ashram, as well.

Varanasi is a quite big and all its small roads leading up and down very often are so narrow that nearly one man can walk through it. Or maybe I was just taking unpopular tourist roads.

I wanted to get on some roof of the house to take a better pictures of temples located at the coast of river Ganga. So very often I went inside anything I could enter in sometimes even finding myself in some apartment!

Architecture is beautiful here. Each house is different and it is almost impossible to tell how old they are. It seems like across the city it is possible to get around in two ways: by roads and by roofs. Roofing is more interesting as I could see a city from top and how new buildings are replacing the old ones.

Got the message from Leo and Margo - they are still in Nepal and as I can't wait for them it will take me a few more days in as I'm leaving tomorrow - will catch up with them probably in some other city.

I'm leaving tomorrow morning at 4:30 AM with the President of ISKCON Varanasi (Boss of Krishna Ashram of Varanasi) to Allahabad where I'm going to stay before continuing further. It's some 6 hours' drive from here and even if I would hitchhike I would not get faster so this is a good option for me.

I got lucky because of a lift from Varanasi to Kanpur as road was so long even driving a car. It is very hard to travel by road in India as roads are mostly like off roads and traffic is slow. Without a lift of a Varanasi President of ISKCON (International Society for Krishna Consciousness).

On our way we stopped at Allahabad ISKCON and picked up two more Hare Krishna devotes towards Kanpur as there is a meeting of ISCKON temple presidents. I'm knocking out in a car all the time as I want to sleep badly. This night was better for sleeping as I got something to scare mosquitoes plugged into electricity. But still I'm sleepy as I waked up at 4AM to get inside this car who totally drops me closer to Rishikesh for about 300 km.

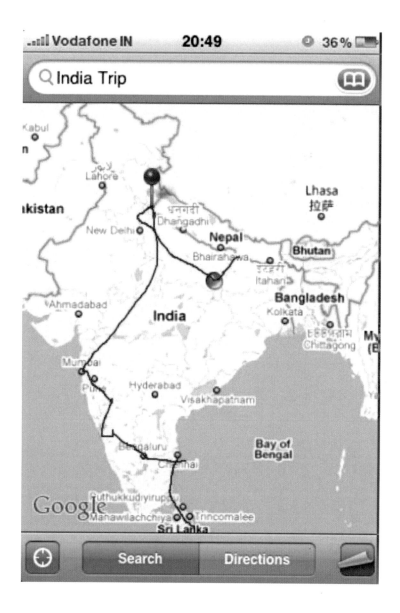

Afterwards I got a roof-top truck and at 9 PM I reached Agra and Taj Mahal. Incredible feeling. The Emperor Castle and Taj Mahal - soooo great!!!!

Yesterday's night I got connected to the INTERNET for free by telling that I don't have a money but I need to contact my friends and send some emails to meet them in Agra. So the internet guy gave the password to connect for free.

So I walked around the Taj Mahal afterwards and planned to start mowing to Rishikesh. As I walked around the great building hugged inside as great walls I heard somebody calling me from the roof top of some building.

This time I looked back because it was not like every "two minutes halo for foreigner" but a different one. The guy from the roof top wanted me to come up and talk with him. And as I did not had any other options - I did so.

He was the owner of hotel and many other things around here and after I explained that I'm traveling without money - he was insisting to stay there for free in his hotel as in ashram as long as I'll want and he even made me a supper. He was telling all the time that money is nothing - it is heart that matters! We were sitting and having some beer. He had a gun for shooting monkeys. And afterwards I went sleep in interesting style room.

So today I waked up at 9 and as I was thinking where to go and what to do I contacted my business partner Vadim. He told me that my travel is pointless without going o special temples in other part of India Tamil Nadu, where in non-tourist temples deep inside a jungles are special and very powerful temples where I have to be on Wednesday. He told me that fare that I can travel anywhere but I have to be there after three days. He told me that as I'm in India he will pay for my visit to temples there if only I can make it until Wednesday morning

As He explained it is important for me to make some magical pujas in there for me. I quite don't understand what all that is about but I felt it might be good and so I accepted his invitation. He will look for a plane ticket for me and probably tomorrow I will fly from New Delhi to Chennai or Bangalore from where I'll have to make 300 km by bus and 20 km by taxi as there where the temple is located no buses are available.

So tomorrow morning at 6AM I'm leaving Agra and beautiful Taj Mahal towards Delhi from where I'm taking flight to Chennai later afternoon. Came back to hotel and listened to the stories about gemstones from my new "guru". And later on went sleeping.

I waked up at 4AM without alarm, took a shower, made up my bag and at 5 there was a rickshaw waiting for me outside of hotel to drop me to train station as my train was leaving at 5:28 from Agra to New Delhi. Took a last look at Taj Mahal and left Agra.

There was no any seat available so I was standing until one man went out and I got his seat in a train and felt asleep. Actually I think that it is possible to drive with no tickets at Indian trains! I waked up in New Delhi and from train station took a rickshaw to the domestic airport. Deam! Delhi rickshaw are charging a great money to foreigners!!!

I arrived at Chennai late in the evening and took a night buss to Kumbakonam where after about 6 hour drive I stayed overnight in a local hotel for 2 EUR per night paying from Tuesdays morning till Wednesday's noon. Generally - except 25 hours non-stop travelling - nothing special today. (From red dot - Agra to Delhi by train to Delhi and from there to purple dot - Chennai - I took the Jetairways flight and from there 300 km by night buss to Kumbakonam. It took me 25 hours)

Traveling with public transport is so boring. But probably in India it is the best way to travel. I have made about 250 flights in my life around 30 countries. And hitchhiking is so much more fun. I had to take a plane only because of timing. I guess I missed the best part of India by traveling from North to South by plane. But maybe I will catch up with it after will finish everything here in Kumbakonam. I might be here around one up to two weeks. The interesting thing is that even if now I'm more South - it feels colder here than in Delhi or Agra.

I feel like tired of traveling today. To many places and impressions in so short period of time. Feels like I am ready to finish this trip soon. Maybe it is like halfway distance crisis. Few days ago my application for Australian visa was declined so I cannot go to Australia and New Zealand. It means that after India I'm going to Sri Lanka where I'll take a plane from Colombo to Kuala Lumpur early October.

That will begin my journeys end as afterwards I'm going to hitchhike from Kuala Lumpur throughout Malaysia, Thailand, Laos and Vietnam to China, Beijing.

I woke up at 4AM without alarm, took a shower, made up my bag and at 5 there was a rickshaw waiting for me outside of hotel to drop me to train station as my train was leaving at 5:28 from Agra to New Delhi. Took a last look at Taj Mahal and left Agra.

(From red dot - Agra to Delhi by train to Delhi and from there to purple dot - Chennai - I took the Jet airways flight and from there 300 km by night buss to Kumbakonam. It took me 25 hours). There was no any seat available so I was standing until one man went out and I got his seat in a train and felt asleep. Actually I think that it is possible to drive with no tickets at Indian trains! I waked up in New Delhi and from train station took a rickshaw to the domestic airport. Deam!, Delhi rickshaw are charging a great money to foreigners!!!

I arrived at Chennai late in the evening and took a night buss to Kumbakonam where after about 6 hour drive I stayed overnight in a local hotel for 2 EUR per night paying from Tuesdays morning till Wednesday's noon. Generally - except 25 hours non-stop travelling - nothing special today.

Traveling with public transport is so boring. But probably in India it is the best way to travel. I have made about 250 flights in my life around 30 countries. And hitchhiking is so much more fun. I had to take a plane only because of timing. I guess I missed the best part of India by traveling from North to South by plane. But maybe I will catch up with it after will finish everything in here.

Kumbakonam. I might be here around one up to two weeks. The interesting thing is that even if now I'm more South - it feels colder here than in Delhi or Agra. I feel like tired of traveling today. To many places and impressions in so short period of time. Feels like I am ready to finish this trip soon. Maybe it is like halfway distance crisis.

Few days ago my application for Australian visa was declined so I cannot go to Australia and New Zeland. It means that after India I'm going to Sri Lanka where I'll take a plane from Colombo to Kuala Lumpur early October.

That will begin my journeys end as afterwards I'm going to hitchhike from Kuala Lumpur throughout Malaysia, Thailand, Laos and Vietnam to China, Beijing.

I was not sleeping all night. Don't know why. I was reading a book then walked around the Kumbakonam, went back to the hotel room, took a shower twice and was waiting for a morning. In a last minute I somehow managed to jump into a bus driving to S.Padhur and it was amazing how just an half hour out from the city absolutely different life was. Kind people, total peace and harmony. I enjoyed it very much.

As I arrived three hours before the arranged time at the temple I walked around the village and enjoyed this so much missed magical moments with nature.

Smiley people, clean cows, happy goats, long coconut palms, forests of banana trees, clean rivers all in stress-less atmosphere - that is exactly what I was longing for last few weeks travelling. It is possible to enjoy India if it's like this: untouched by "hello" to tourist each 5 seconds, peaceful and quite with squirrels squirreling around.

At Ten I had a meeting arranged by my business partner Vadim in the temple where local rishis will do the homa on behalf of me and Vadim. Homa is something like a ritual made to remove spiritual obstacles and grant particular person with blessing from specific God's energy. Today I have two homas - each in different temples and for different purposes. One of them started at 10 in the morning and the second at 15:30. So it's a full day for me today.

Basically my ascetic expedition through Eurasia ends today as this was a reason I was going to India through all Russia, China, Tibet and Nepal. This was my pilgrimage. I will stay few more days in Kumbakonam to visit some more temples and further my "ascetic expedition" is turning into just a travel.

From today I'm a tourist but still continuing tracking and exploring Eurasia's second part - Asia: India, Sri Lanka, Malaysia and Thailand. But this will be in a different way - this is already way to my new home - China - to where I will move after spending at least a week in Goa and Sri Lanka beaches where I'm going very soon. Why? Because today I did everything I could not do in the Mt.Kailash - the main reason for my trip. I did it today. With a help of my business partner Vadim, few Brahmins and a much hidden temple where tourists are rare. I'm done!

My ascetic expedition is over. I got what I came for almost in two month duration. Later afternoon I hitchhiked to another temple and where I had another homa for different purpose.

It ended only at 7:30 PM and was quite magical as there was a need of participating young children in it - there where three or four of them - all sons of Brahmins (priests) and they where taught for particular purpose to chant Vedic mantras together with a priest. Quite fascinating and spectacular that was! I was home only before nine PM; full of impressions, emotions and memories. Went sleep almost the same time as entered - two homas per one day that is a lot of work. I was really tired..

When I left Europe, my friend Dr. Eric Beeth from Brussels, told me that I should not be worried that I'm going without money in this trip. As he told according to his experience: 1) a trip like this usually brings more money in and 2) real starts when money ends. Today is a first day of my new life.

At least this is what I'm telling to myself and what I believe in. I was wearing all my rings today I bought in Nepal: beautiful small pearl, dark brown smoky topaz, small blue sapphire, not very clear and shiny, but still - beautiful dark red ruby and my favorite - amazing emerald in green color. And I have to admit - the Doc. was so right about what he said! All these stones I would probably never bought in Europe -and most of all - wear them. But here - I can wear them and feel comfortable with them. It's not about the stones - it's about the qualities they contain and pass to their owner.

Different stone has a different character. And they are acting like a small friends and if used in a right way. I know a doctor who is even healing people from HIV using his own method of healing with precious and semi-precious stones. So I've gained stones from my trip!

I waked up early today and went to the same temple I was into yesterday to continue magic rituals called homas. Basically I spent whole day there as when I came back I was to tired to do anything else.

I went to the internet cafe and was sorting out videos and pictures from the trip and uploading some of them on picasa and YouTube. So now it is possible to view some more videos from trip at my YouTube profile: youtube.com/martinsate

As I'm making my notes into the phone and there are some problems with wi-fi in this city I have not uploaded any posts to my blog, but I'm really hoping to do it soon.

Generally nice day. The rest of the evening I spent on preparing myself for a China where I'm going in about one month. And then just knocked out in a bed...

Today I waked up at 6 AM to go to the temples. Here is so many temples that to visit all of them would take about one month minimum. So As I'm planning to leave tomorrow and afternoon I have one more homa Vadim organized - I had to get up and rush to the temples!

I visited about 6 of them today. All of them amazing and in each of them asked to do "archana" for my family, my friends and friends families. Archana is ritual made by a priest. Kind of prayer on behalf of the person who asks for it.

It is incredible feeling to be in the places I was reading about in Mahabharatha, Ramayana and other Hindu religious literature and to know that since the very that time when the story was made-up there is a temple standing in that place where for thousands of years people are coming from all over India to remember that event.

Also I got a lift by the elephant today! that was amazing! I enjoyed.

I fed him with coconuts, bananas and flowers for his amazing work! he dropped me almost until the temple where Son was Preceptor to his own Father - Swaminathaswamy.

Totally I drove today about 100 Km around all the temples. After each archana, puja or homa there is food given by the temple - called "prasad" - meaning - food offered to God. So for last few days - since I'm here and going around the temples - I've been eating only this type of food. In the temples, with hands.. Rice mostly.

Later afternoon I went to my last homa on my way visiting one more temple - 30 km away from my hotel and afterwards again to the Shukra Temple for my last homa.

I'm leaving Kumbakonam tomorrow morning at six. So as I will be on a road to Goa - no news till I'll reach it.

Leaded by a will to see one of the biggest Christian churches in India I drove to the Velankanni where the church is located at the sea shore.

I left Kumbakonam at 6 in the morning and arrived at Velankanni only around 9 so the journey took about three hours.

After I arrived and passed through tens and tens pf beggars I finally reached the white cathedral. It was a weird feeling that Hindus are going and praying in it but as I spoke with few locals I found out that their religious believes are like Jesus was a Saint like other Gods. Basically - yogi - who have mastered his mind and used it towards other being happiness.

Afterwards I found a wi-fi and spent few hours on the internet and went to the beach where I located small and cozy temple built for Shiva.

I did not swam in the sea even if I wanted to because I did not had any place around to put my belongings and there were too many people around. But I had a fresh coconut juice on the beach!

Later I found some peaceful place to sit down and enjoy the amazing music of sea waves beating as they were coming to the shore.

Today at 6 pm I'm leaving from here to Bangalore from where I will take a train to Goa. In my thoughts I'm there already. This Bay of Bengal reminds me how much I missed the sea this summer and how badly I wanted to go to Goa to catch some Sun and lay down in a beach for hours. Can't wait Monday when I'll arrive there.

Soooooooooo!!!!!!!!!!!!! Beautiful!!!!!!!!!!!!! First city in India until now I can say for sure - I enjoyed it. I mean - all the cities I've been to was because of temples - like Varanasi, Kumbakonam... it was all to be in particular places not to enjoy the city. And the rest of the cities I went to was because of I just traveled through them.

Even if Bengaluru was just a transit city for me I enjoyed those nine hours I was there for 100%! Firwt of all people are more used to tourists and are not behaving like in other places when they see "a white man".

I walked about 15 kilometers. Visited amazing Bengaluru Botanical garden, Bengaluru Palace, park, where everybody is throwing out belongings from pujas and archanas and other many places that got me astonished.

It is rare when inhabited places in India are truly enjoyable as because of noise cars are making - everybody is beeping each five seconds making a lot of noises, people always are shouting loud "hello's" and shop owners wants to sell some unneeded stuff after each ten meters.

My train towards Goa was leaving at 15:15. I found a station by GPS in my phone, had some food on my way for ten rupees (about 15 Euro cents) and stepped inside a train leading to the Paradise on Earth - Goa. This time I had a sleeper seat - I paid less than 4 Eur for it - for 700 kilometers and 14 hour drive.

So today was a really relaxing day comparing to others. I was sleeping in the park, in the train and generally felt rested.

Yeahhh!!! at 6 am I arrived in Goa, took the cheap taxi that dropped me 45 km till the hotel I'm staying in for 4 euro and also I rented a scooter to drive around it as the transportation is understandable for me here.

To rent a scooter for one week I paid 20 Eur. My hotel is 20 Eur for one week, too. But the food is expensive here comparing to the rest of India.

I love this place! It's amazing!!! It is the best place where to celebrate two month of my travelling! The hotel is nice and clean and there is one waitress when I noticed her I could not take my eyes off of her.. She is amazing! Just can't remember her name...

Goa is beautiful. I'm sure I'll enjoy my next ten days here. Relax. Chill. Sunbath. All in one at the coast of Arabic sea. I'm high just to be here!

Yesterday was a good day. After celebrating my two month travelling at some pub on the beach I was so tired that drove to my hotel room and knocked out for 14 hours until today.

This day started with a scooter drive to Old Goa. It is so beautiful and peaceful here! I found out that it is possible to get a property with amazing sea view for only 15,000 Eur!!

I visited churches and temples. Basically was just driving around and enjoyed this amazing place.

Later afternoon I gave a flowers to the girl who works here in the hotel as she is very beautiful. She was surprised and showed her beautiful smile to me.

(Ganesh Temple - had a puja and archana, soon will have homa here). During a day time I took a nap. Evening I spent at Anjunas beach where all the pubs on the beach are located.

So nothing special for today - just enjoying Goa...

(the door behind my back is my hotel room - right on the beach!)

Nothing special for today. Spent all day on the beach with some Russian girls..

They were driving me around as there were moments I could not do it myself because of the reason I'm in Goa.

The same reason I have not been till internet for three days...

Just few things in Goa are very good...

Like the Sunsets at Curlie's beach bar...

Today I went out even without a phone with me as I'm prepared to make some changes in my daily Goa routine and go to shaman White Dove who will treat my chakras.

So afternoon I spent at the beach, again, doing the same things I'm doing every day.

Eating. Sleeping. Smoking.

This day started differently as in the very early morning I had a Ganesha Homa 35 km from place I'm living in.

So after the homa I went back home only about noon and decided to go down till net cafe.

While I can...

And afterwards I spent all day by PC doing some jobs..

Girls wanted to have homa for them at the same temple where I had it so I brought them there.

They waked me up at around 11 and after a quick shower we moved towards Old Goa where after 20 km a temple I had a homa was.

It was a fun driving. We took a lot of pictures on our way and had a lot of fun.

Afternoon I spent at Curlies and Shore Bar. Had some food and later went back home.

I have got everything I came for here in Goa. Peace. Harmony. Place to charge up the batteries with crazy sunsets!!!

Today is my moving day! I'm moving even closer to the beach and closer to the parties! So I had my bag on a top and early morning drive to my new hotel. It's amazing!

Feels like the New Life started just now. Met some nice Japanese girls... have some swimming with them later afternoon...

Yesterday everybody was enjoying Goa's atmosphere at my place. Around 10-15 people came. We were sitting down in a terrace at the place I'm renting and enjoying sounds of the sea waves.

Then the owner of this hotel joined us and as we talked about a lot of things somehow we got until a website, too.

I'm not actually now up for making anything but somehow it might be good for a lots of reasons to make one for him as I really like this place and secondly I need some cash from time to time so this website might cover my further costs of travelling.

So today I waked up early. Without putting on any closes except swimming shirts I stepped out of my room to the beach and had a morning swim in the sea.

Then I waited for my new client and got back in work!

Today is 70 days since I left cozy life in Brussels and hit the road towards where I'm now. 10 weeks!!! So today I decided to make a party!

I got into good relationships with the hotel owner I'm staying in Prasad as I was making a website for him so I talked with him regarding a possibility to make a party at the place I'm staying with a reggae music what he was happy about.

So today evening time I'll have some reggae party at my place. While everything is preparing I finished the website - you can check it out here www.saiprasadgoa.com

Otherwise I'm still enjoying Goa's atmosphere and riding a two-wheeler for boys - I got almost brand new in astonishing purple blue color!

So more pictures next time!

Bob Marley was shouting on all the beach and I was lightning the candles with number 70 - meaning 70 days since I'm travelling.

Yesterday my reggae party was attended only by few: only by Russian girls I've been having pleasure to enjoy Goa for a recent week. But still it was good. Especially when the Sun was going down.

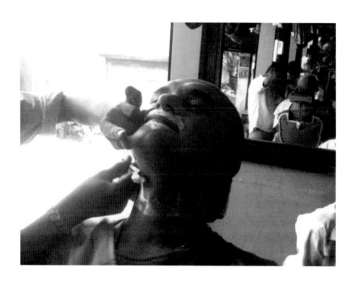

I cut some coconuts off the tree and made a fresh juice from them. Video is on a youtube.com/martinsate

I made some money yesterday for making a website. So I'm going to leave soon, I guess. Somewhere further - let's see where the road will lead me...

Today's morning started with giving back my bike as I decided to spend the last day in Goa by enjoying stillness and peacefulness at my hotel - as it is sooo good - right at the sea! I'm going sleep and waking up with a sounds of waves. It is amazing.

So - still enjoying Goa! For a last day.

In the morning at 12:00 taxi was waiting outside of the hotel to take me to the airport. On my way to it I grabbed a girl who was going same direction and I took my flight from Goa to Chennai through Bangalore.

Yesterday kind of a leaving party turned out at my place when few of my friends came. Even White Dowe was at the party and afterwards even gave me some healing.

I enjoyed my last day on a motorbike as I finally decided to have it for one more day - the last day in Goa - to completely enjoy it: I've noticed I've become better in driving it. Also one Canadian Princess I met few days ago and I was looking for her for last two days - I finally met her accidentally 30 minutes before she was leaving Goa. We quickly exchanged emails and I drove her a little bit around the beach area at the back of my bike. As I was having feelings like sympathy towards her - it was a real pleasure to have her behind me.

Afterwards I had a two flights - one from Goa to Bangalore and second from Bangalore to Chennai with Kingfisher airlines - the same ones I'm drinking a beer from... Tomorrow I have a flight to Colombo from Chennai. So today I'm staying in Chennai.

In Chennai I got some cheap room for 120 rupees (about 2 Euros). Tomorrow will see... continuing towards Sri Lanka.

Since the very morning I've been travelling: to airport in the morning at Chennai and now from Colombo airport to train station to get me deeper inside a Sri Lanka. I'm amazed to be here.

Flight was short - only 1,5 hours. More news will write later as till now it was only airplanes and other transportation which is not worth of writing about.

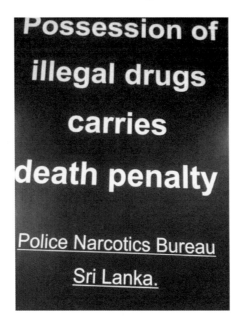

It seems like I'll have a break from some things in Sri Lanka. Good for My health!

From Colombo I took a night train deeper inside the Sri Lanka towards place where the Buddhism was first introduced to the King of Lanka a lot of centuries ago - Anuradhapura.

When I opened my eyes - it was 3:30 and I just missed my train station and train was just about to take his speed. I quickly grabbed my bag and jumped out of it as the doors was opened. I had downloaded a map where I need to go. It is called a sacred city.

As a lot of temples and other stuff is packed here since a long time ago.

Here is a big statue of Buddha, temple built on a place where local King a while ago was taking a bath and his scepter disappeared. So as he could not found it - he had to build a huge stone monument around it so that others don't find it, too...

I went to guest house called Boa Vista in about 7 am and found out that a room price is higher than I was expecting. The cheapest room I could get around 15 Euros - but for me it is too expensive. So I switched on the "Plan B" by offering a website and shortly - I had a huge room with a lake view, dinner, coke, coffee, tea and everything I wanted including internet - because I got them as my clients!

We continued this morning by discussing what needs to be done in a website and around 10 AM I went sleep till late afternoon when I after a small chat with the owner Shanida started to make a website.

As I'm staying and eating here for free till at least Tuesday - it's about 4 nights - for the website I'll get only 100 USD, which is OK as I like this place, owners and atmosphere.

So today and tomorrow will continue my work on a website and afterwards will see what is what and what this prehistorical island can show me!

For a last few days I was thinking about "ascetic" expedition as it is stated in my website. There is written that I will avoid paid transportation and hotels. But recently I was staying in a quite expensive hotels in Goa and now in here. Also I've used airplanes a little bit recently. So how to explain that?!

Here is a point that when I reached a Goa - i wrote that my "ascetic expedition" is over. Now I'm just a traveler. But still - even if it is not so extreme as it was in the beginning - the ascetic now means to continue everyday travelling with its duties. I have the same bills to pay as anyone - food, water, transportation, phone, shelter; only I'm getting a cash for it on my way

- which makes it more difficult - that's why I feel still continuing my "ascetic expedition" even if I allowed myself to have a great time for the last few weeks.

Secondly - I'm doing some duties: travelling is not easy, especially when it's so continuous. Also I need to write something every day. And to write about something - I need to see something...

So I stayed home today all day and as the only costumer felt as a king in a castle because of servants who were very kind to everything I was asking for while I was making a website.

I made some changes to my own website. Mainly - took of the big "ascetic" from the front page. So you can write me some reviews about it...

I waked up around 12 today or even later and after having my breakfast the servants made for me at the outside terrace I went to look at the city.

Firstly I went to the Royal Pleasure Gardens which were just across the street by the lake. I felt such a peace in there that I even was meditating for about an hour. And felt absolutely blissful state.

I felt like I would live here in a previous lifetimes. This place seems so familiar and lovingly. I feel like home here.

Afterwards I went down to see the tree where first Buddha got enlightenment couple of thousands years ago. Yes - that tree still exists and is big. Surrounded by guards and buildings over 10 meters high to assure nobody touches it. Also there is a Buddhist temple by it.

And later I was just enjoying everything what this city offers - mostly sightseeing and when it became dark - I went home. Had my supper and was working with few websites.

I'm waiting for a hotel owner to return from Colombo where she went 3 days ago to pick up her camera to take a good pictures for a website. Bit she told that she'll be back after one day - but now it's been already a while since I haven't seen her.

That makes me kind of stuck in here and Cutting of my days to see Sri Lanka. But from other side - I'm doing things which had to be done a long time ago. And I'm spending all days by computer with tea brought to me by my servants for a time I'm in Sri Lanka.

Today since the very time I waked up I've been working on one of my books to be published. So it's a much more time consuming process that I've imagined that. But it's moving.

And I guess I'm lucky to have the opportunity to work on a book in such a nice atmosphere - a big house by the lake, surrounded by palms, nice weather and bird songs wherever I go. I have not been out of this castle at all today. It does satisfy my every need. The main thing - I'm a King here for these days. Well - at least I feel so. So as I am what I feel - I'm a King! Yeeeeee! Don't know yet when I will leave this magical place. I believe that I was living somewhere around here few past lives ago!

Finally!!! I made it today! I finished my work on a website for this hotel I'm staying! it was a tough work which asked me for almost four full working days. For a website like that I would usually charge a minimum 1,5k Eur, but now they got it really cheap as I'm staying in here and actually...

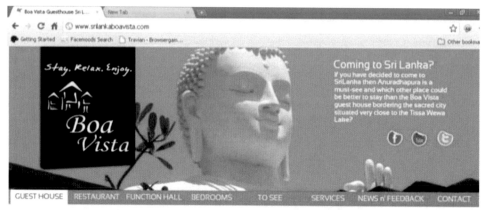

I feel like sold myself cheap. But anyways... Whats done - done. I even shoot a video for them. That's nice. I like it, too. You can check the site at www.srilankaboavista.com so generally - that's it for today - nothing much more.

Probably I'm gonna leave tomorrow morning for my next journeys. Enough of Anuradhapura even of it's nice here. And yes: I do recommend this guest house for everybody who comes this side. It is truly splendid views and amazing house!!!

So today I'm leaving from point - A - Anuradhapura – Sacred City, where I stayed almost one week in a Boa Vista hotel where I must admit - I was treated like a King with all those incredible views and walks in Royal Pleasure gardens which were just across the street and I used to go there for to call to my friends and family as because of amazing atmosphere in there.

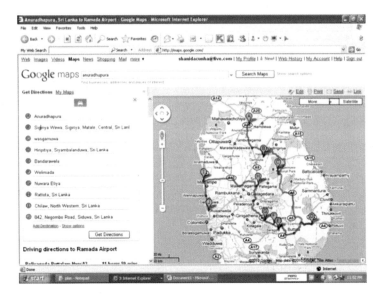

So today I'm going to my next city - Dunuwila, where Lord Rama fired the Brahmaastharam (something like a bomb just 10 thousand years ago) at King Ravana (who was a King of Sri Lanka and 10 more countries that time), which eventually killed him. Dhunu means arrow and Vila Means Lake. This place got its name because Lord Rama fired his arrow from this lake. But this arrow reached Ravana at the other side of Sri Lanka.

And I will share my further plans which are as follows and according to the points in a google map:

B. Sigiriya Wewa. Rock garden. Asia's oldest surviving landscape garden. It is also believed flamboyant Kasyapa aspired to be The god king with the Sigiriya palace as the very personification of his divinity, to rule his kingdom up high, like the god king Kuvera up on the Mount Kailash of Hindu mythology.

C. Wasgamuwa. Dunuwilla.

D. Hiripitiya, Dolu Kanda. Herbs dropped by Hanuman for healing of Rama when he was unconscious after Ravana hit him.

E. Bandarawela - Ravana Cave & Tunnel Network This Buddhist shrine at Kalutara

F. Welimada – Sita / Ravan Caves – Ishtripura. This was one of the places to which King Ravana shifted Sitadevi as a precautionary measure which he was forced to take by Lord Hanuman's advent. There are lots of intruding tunnels and caves in this area. This seems to be a part of a great ingenious network of paths, which is interconnected to all the major areas of King Ravana's city. Sitadevi took bath in this very stream and had dried her hair sitting on a rock and put clips to her hair, hence this rock is known as Konda Kattu Gala. This is situated in the Welimada Area.

G. Nuwara Eliya - Ashok Vatika is a garden where Rāvana held Sita captive. This is in the area of Sita Eliya, close to the city of Nuwara Eliya. The Hakgala Gardens located at the base of the Hakgala Rock forms part of the famed Ashok Vatika. The Sita Pokuna is a barren area atop the Hakgala Rock Jungle where Sita was kept captive. Sita Devi is set to have bathed in a stream at Sita Eliya.

The Sita Amman Temple is located at this spot. The stream that runs from the hill, catered to the needs of Sitadevi during her stay at Ashok Vatika. She is said to have bathed in this stream. About a century ago of three idols were discovered in the stream, one of which was that of Sita. It is believed that the idols have been worshipped at this spot for centuries.

Now there is temple for Lord Rama, Sitadevi, Laxshmana, and Hanuman by the side of this stream. It is interesting to note that foot prints akin to Lord Hanuman's are found by this river and some are of small size and some are of large size, which tells us of the immense powers of Hanuman transforming himself into any size.

H. Rattota - Rama Temple. There are a few Rama temples in Sri Lanka, this is one of them.

I. Chilaw. Munneshwaram Temple. Temple dedicated to Lord Shiva was located here before Rama came. God Shiva blessed Lord Rama here and advised installing and praying for four lingams at Manavari, Thiru Koneshwaram, Thiru Ketheshwaram and Rameshwaram in India, as the only remedy to get rid of the dosham. The first Lingam was installed at Manavari about 5 Km from here, near the banks of Deduru Oya. I hope it won't take too long!

I spent morning by saying warm bye-bye to the great host and so nice person Boa Vista's guest house owner Shani. It was so nice to meet her and it was so amazing how she and her team treated me. I want to go back there some day for sure.

After breakfast at around 12 she dropped me till the bus station which led me further down my track and around 3 PM I reached Dambulla. I visited Golden temple and Rock temple which was just behind the Golden temple. It was a long way up through monkey companies.

Afterwards I went to Sigiriya to the old Sigiriya Rock kingdom. It is told that this place may have been inhabited through prehistoric times.

It was used as a rock-shelter mountain monastery from about the 5th century BC, with caves, garden and palace which were built by King Kasyapa.

Following King Kasyapa's death, it was again a monastery complex up to about the 14th century, after which it was abandoned.

Entrance tickets are enormously high in price in here. Like to visit that rock temple costs about 8 euros but Sigiriya - about 15. I think it's to expensive! Well but anyways - once I'm here - I had to go there.

So tomorrow I'm leaving at 6AM towards my next checkpoints - passing few of them without stopping.

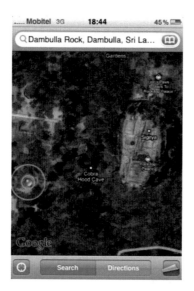

(I was there on a top)

I met a girl from N.Y. on my way to this place so she told me about some cheap hotels she read about in Lonely Planet guide book (G., I even didn't know about existence of such a book until I was in Tibet!) so I stayed a night there.

We had some beer and afterwards I went sleep in my bed covered by mosquito net to keep me safe from many diseases mosquitoes are happily giving away for free in this area, including malaria.

It is so imaginable incredible fantastic!!! This is a place worth visiting Sri Lanka!!!!!!!

To get here I got up at 5:40 and left with a bus to Kandy, from where I went further down to Nuwara Eliya because of a time matter not stopping in a places of less importance than Nuwara Eliya. This is a significant place and I'm happy to be here now as it is the 80 days of my travelling and there could not be a better place to be.

Nuwara Eliya or Ashok Vatika is a garden where Ravana held Sita captive. This is in the area of Sita Eliya, close to the city of Nuwara Eliya. The Hakgala Gardens located at the base of the Hakgala Rock forms part of the famed Ashok Vatika. The Sita Pokuna is a barren area atop the Hakgala Rock Jungle where Sita was kept.

Now Sita Amman Temple is located at this spot. The stream that runs from the hill, catered to the needs of Sitadevi during her stay at Ashok Vatika. She is said to have bathed in this stream.

There is also temple for Lord Rama, Sitadevi, Laxshmana, and Hanuman by the side of this stream. I sit down in these incredible valleys and kind of was feeling all that through myself. I'm really impressed by the Ramayana story, it has captured my heart for sure a long time ago so this is a place where I believe one of the greatest pains of Love and sufferings where taking a place as Sitas Love towards Rama was something more than just Love. It was a very pure and uplifted love, as both of them where very pure and uplifted personalities.

It's a mountaineer area here, about 2000 m beyond sea level and It's quite cold so I even had to put on my jacket. Today also I visited one part of an underground cave labyrinth of King Ravana built about ten thousand years ago, where that time he was keeping Sita after Rama's assistant in the search for Sita - Hanuman found her in Nuwara Eliya area.

It is amazing to be on the sites and see everything I was reading about a while ago and it seemed like a fairy tail until now when I actually can be there and see the things written about thousands of years ago in life. That's amazing experience especially if the trail I'm tracking is as important for Hindu people values and religion system as visiting Israel and Palestine for

Christians. I do respect both of them it's just that story of Ramayana is so incredible and it shows so good a real values of life in a simple way so I'm touched to be here. By the way - I've been to Israel and Palestine, too. The feelings where similar. Those who have been, too, probably will understand how I feel.

Afterwards I took a night train towards side of Colombo where about a hundred kilometers North a temple is located in which Rama, after killing the king of Lanka - Ravana, about 10,000 years ago, asked Shiva to free him from a sin of killing as he felt some bad energy following him. This is believed to be a place where the first temple for Shiva was standing. I got there at night and spent all night waiting for a morning for a priest to come and make a puja. So that's it for Sri Lanka! I had incredible time in here and this is a really nice country to visit.

I hearty valued the opportunity to be on a Hindu's Ramayana trail, Buddhist Sacred City Anuradhapura, where the oldest tree in a world is located under which first Buddha got enlightened and also can't mention kingly spent time at Boa Vista hotel there. So I'm ready to go further to the city whose name was always intriguing me since I ever heard it for a first time - Kuala Lumpur in Malaysia. I'm going there NOW! Incredible!

Nice. I like the name Kuala Lumpur so much. It sounds like something really exotic! Like advertisements for Malaysia- truly Asia. I came to airport in Colombo much before I could even check-in as I was going somewhere from a middle of Sri-Lanka and in Colombo I arrived very late so I decided that there is no reason of taking a hotel for few hours and I went directly to the airport.

This time I'm trying kind of new company for me - I've never used them before - Air Asia. They were the cheapest - flight from Colombo to Kuala Lumpur was about 150 Eur or even less! And as I was sitting in the first row - I got a seat by one billionaire from Netherlands who owns a set of companies around Asia doing agricultural stuff. We had an interesting conversation and fun - we even pretended that we are sleeping when we took a pictures as for both of us this was a first time flying this airline company.

My check-in time started at 5:15 am. So early. So I wasn't sleeping all night. Was looking on a map to see what to see in Kuala Lumpur and was thinking how long will I stay there. Also I planned my trip further to Thailand which is very close comparing to the distances I've made so far. I'm even thinking about hitchhiking there - must be nice - I already miss that part! As last few weeks I enjoyed Goa and Anuradhapura - the places which are not about to call "ascetic" expedition at all! But generally - I'm still holding on my plan - no outside funds have been involved except those needed for my flight from Delhi to Goa which was a matter of time not my choice as I had to be there on a homas. So - so far I've been strictly holding to the principle - earn and travel and earn as its stated in a websites "About" section!

From blue dot to green I took a flight today. From green to red I'm going to hitchhike for next days. So - Kuala Lumpur! I've landed at afternoon and went out from the airport just before 3 PM local time as during my way I was changing time zones, too.

The city itself is amazing! It's amazing! Very beautiful place!!! So many ancient buildings and everything is so clean and tidy not to mention fascinating architecture including Petronas buildings - twin towers, which I reached by quite a late time. So as it was late but there was a lot of things I wanted still to see in this city and I had no proper sleep for already nights - I finally decided to pop in some hostel I found on my way.

Knocked out in a waiting mood to see Kuala Lumpur tomorrow! It's fascinating city!!! ...and the name....

Ouuu, last 24 hours where crazy! I even don't know from where to start my story... Maybe pictures will tell everything.

(This is a picture I don'r remember taking...)

First - I got drunk in the Thai Bar. Secondly - when I waked up - there was a girl sleeping next to me. Thirdly - I can't remember half of last 24 hours..

So I was sleeping all day today and got out only at night. In almost "autopilot" I went to the huge television tower and went to the top of it.

Afterwards just continued clubbing. So - nothing special.

Not so much up to writing today...

I even went to the temple

After last nights farewell party in Kuala Lumpur Reggae Club I was home around 5 a.m. - making progress as previously I came only around 9 am... I waked up with a wake up call that I need to check out from a hotel.

Quick look in a mirror, red eyes... I guess to much party - at least this time I waked up in my room and alone. Took a shower and after moved through a city towards bus station to go to Bata Caves. Amazing place with place to worship Hinduism gods inside. Monkeys, of course... One was even biting me in my leg. That was funny.

(huge statue of the one of the Shiva's sons)

And then I went back to the Kuala Lumpur which is about 15 km from there and bought a train ticket towards Pulau Pinang Island, about 350 km from Kuala Lumpur towards Thailand where I arrived at 11 pm only in the train station called Butterworth, from there boat to the island passing nearby magnificent around 15 kilometers long Pulau Penang bridge and that's it - I'm in Georgetown - island.

I met one guy from Czech Republic so it was easier to travel and find places. We found some cheap hostel and went to bed.

No party today.

Yesterday's night I could not sleep so I went down to the internet cafe and was working till 4 am by the computer. So many things to do - sometimes I'm so happy that I have a time to do the things I have to do even if I'm travelling to get t some cash for further journey.

So today I waked up about 11 and at 12 there was a checkout. As the Czech guy is going to Thailand, too, both of us are going to travel together for a few more days untill our road leads the same way. So we went to the ferry and left the island towards train to Thailand.

For a few euros we got on the train towards Thailand and it was standing for about two hours and not moving just before the border. Everyone was pissed.

I'm staying somewhere in Thailand right after the border today. Nothing special - just travelling today... However it was not so smooth yesterday as I told in my previous post. My hopes of staying right after the border almost got vanished away with a big risk of not getting across the border. So all the story in details would be like this: The train we were driving broke down just about 20 kilometers before Pedang Besa Malaysia - Thailand borderline. After waiting in the train for about four hours we decided to hit the road by hitchhiking. So at 8 pm - when it was dark already - we finally decided to leave the train and went towards highway to the border. It was raining heavily.

When we were picked up by a guy who dropped us until the border - he explained that actually we are in the last minute as border was closing at 10:00 pm and we had only half hour left. So we were really rushing. BUT: After Malaysian immigration let us pass theirs border - Thailand didn't let me in. They allowed to cross the border only the Czech guy as he had a visa but I didn't. Unfortunately this border crossing point is to small and they don't provide visas on arrival service. So the Czech guy continued without me but I had to go to the other border crossing point which is about 70 kilometers. And there is no bus. And there is no train. And there is no any kind of public transportation except taxis who charged a noticeable amount of money so I had to deny their offer and got back to the "basics" and was hitchhiking again towards Bukit Kayu Hitam.

On my way I found out that I don't have much time left as the other border was closing in about one and half hour - at midnight. It was raining very heavily and I was almost pleasing the cars to pick me up as there was going average one car in ten minutes by that evening time.

But finally one car stopped and the driver was going to Thailand, too, so he dropped me right until the border in about 20 minutes before it would close. After I got my single entrance kind of visa for 15 days for free I stepped into the Thailand. It was midnight. No trains. No buses. Expensive hotels. Sluts. And as dirty as in India. And I felt like just went back from the paradise to hell. And as there where not so much choices I decided to get somehow towards next big city - Hat Yai.

One lorry driver picked me up and after three hours by paying about 2 euros and, by watching the driver drinking whiskey from time to time until one bottle was empty I got there and spent a night at the internet cafe working until the morning from where at 6:30 in the morning I took a bus to Krabi.

So now I'm in Krabi!

Wow - after around 6 hour drive in the bus I finally arrived at Krabi. It's raining in here a lot so not so much as a paradise even if it's almost at Phuket - the top tourist destination in Thailand.

I went here because of Tiger Cave temple - the old temple built on the rock with 1237 steps to climb on top of it which I eventually did. But sometimes people are too scared and they don't.

Afternoon I spent driving to the Phi Phi Island which is about 45 kilometers inside the bay from Phuket.

And on island I found some cheap but nice place to stay and met some Russians and spent all evening with them.

Today I was just walking around the island, swimming, relaxing, chilling, doing nothing and went to the Thai massage.

My new home

Chillin'...

At the evening I could not sleep and I decided to go out for a beer at the bar in front of my hotel. Reggae bar is doing kind of funny fighting and I decided to participating..

My opponent was a Swedish guy and have been training taekwondo for five years... of course he won but that was more for fun not for winning or losing.

And afterwards I just went on the beach and was hanging out with some Thai girls...

I waked up around twelve today and went to the pool instead of shower. Jumped in, refreshed myself after yesterday. There is a free booze at almost all beach bars after 11 pm so everybody gets drunk.

I spent all day laying by the beach at the hotels terrace, swimming in the pool and hanging out with some quite pretty and nice Russian girls from Moscow as they were at the same pool where I was.

Afternoon passed up in the Phi Phi island view point just by sitting and enjoying warm breeze of fresh sea wind watching the palm tree Mountains and the beach. By the way- movie The Beach was made around here!

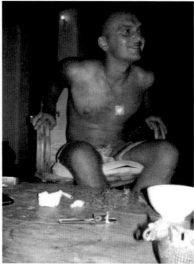

Yesterday was such a beautiful day. At evening time I met a Russian girls and one of them is so very lovely that I almost felt in love with her. Rusalina. I could not let her go for until 5 am when I just felt having too much of everything what Bob Marley suggest and I just decided to take off towards my bed.

We were just chilling on the beach, watching fire shows and dancing all night long.

And today I had to wake up at 8 to go to Maya bay - where movie "The Beach" was made. I was snorkeling about an hour there, and having some beer at the beach.

Later afternoon I decided to go away from this island as everybody I knew was leaving including those Russian girls. So at 2 pm I took a boat to Krabi from where bus to Bangkok which goes there about 15 hours.

I was mostly sleeping all day today after yesterday's long ride from the Phi Phi islands. Just went to few temples and went back to sleep.

It's actually not so bad city. I was expecting worst. Even if it's massive and huge and high and big somehow I find it cozy.

People are very nice and polite. You can buy a diplomas and driver's license from all around the world here. So if you have any queries let me know: D ...

O, and today is a 90 days since I'm travelling - will celebrate it in kind of special way! Bangkok, private style...

Yesterday I had a chat with my friend from Latvia who asked me when I will finish this trip. And I become thoughtful. Yes. I'm done. I've seen everything I wanted to see and so much more that it would be a nice time to end it here and now.

So I'm thinking about going home. I waked up early in the morning, packed my stuff and left for a bus going north from Bangkok to Ayutthaya where the ancient Thai civilization has built a lots of temples. I guess that is my last destination in Thailand for this time. But it is definitely the place where I want to return.

Ayutthaya impressed me. The temples and stupas where amazing. And they are found all around the city. Big and small old and older.

In the afternoon I went to take a train to the North of Thailand to enter Laos, but because of rain recently the part of the territory is flooded and I had to make a little changes in my plan entering Laos. The good thing is that anyway - you can get a visa on almost every border so my changes where even better for me.

I'm entering Laos. And yes: I'm going home. Today are the three month since I'm travelling and it is enough for me till now. I am tired, really.

Yes. I have that relief I had always when I was leaving something. Leaving jobs, homes, places, girlfriends, situations. The relief. And I got it now leaving this trip behind. It means - now's the time. The time to END. To end this journey. Today I met my wife in the bus. Will go back home and offer her marriage.

That is so funny - I even don't know what to do with a domain name vedatrac.com; what to do with a space I'm renting for a hosting... And what to do with a blog I'm writing - it has become like a diary to me - a daily thing to do no matter how I feel: kind of meditation.

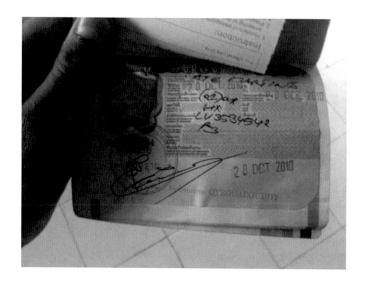

I think I've got what I was coming for in this trip. My own resume - I did good; not excellent which would mean sleeping every night in forest and hunting squirrels for a food. But I'm satisfied.

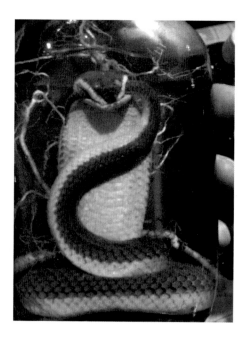

This trip definitely has changed my life. And it's the best time to end it. I'm really tired of it, too. As I stated previously - I was done just before the Goa - from where all the chilling stuff started from and "ascetic" expedition ended.

So from now on - I'm done both or any kind of expeditions - I'm going home. Even if I'm a "rolling stone" and my home is "wherever I lay". Now I'm going home. From today. So ... Few more days of this blog. Few more days of this website. Few more days of the daily news. I'm really done. It's finish.

I'm going home. And it feels much better than just good! Thank you everybody who was following my trip !and sharing emotions with me. Thank you all those people who helped me on my way. Thank you everybody who was interesting in this and gave me a moral support to continue my travelling. And thanks to everyone I just met on my road and had a great time with!

I'm writing Laos as I just can't understand the cities I've been and going - all of them seems to me the same even the names. Or maybe my mood is out of travelling so I don't notice the difference which probably exists. But there is one thing I've noticed since the very first day I entered Laos - no Wi-Fi. And for me currently it's enough good reason to leave this country ASAP.

It's a good country. Really, I mean it. It's a good for relaxing, meditation, doing nothing and being with a nature. But the options offered here for tourists are so limited: get drunk with a local cheap whiskey, get drunk with a Beerlao beer, get drunk while floating stoned and/or under mushrooms in the river, get drunk in the... Generally - get drunk anywhere and party like an animal. Probably that is a reason why this place is loved by tourists in age until 25.

Wherever I go here - it's nothing. Probably it's because massive aerial bombardment was carried out by the United States during the war with Vietnam. The Guardian reported that Laos was hit by an average of one B-52 bomb load every eight minutes, 24 hours a day, between 1964 and 1973!!! US bombers dropped more ordnance on Laos in this period than was dropped during the whole of the Second World War. Of the 260 million bombs that rained down, particularly on Xiangkhouang Province on the Plain of Jars, 80 million failed to explode, leaving a deadly legacy. It holds the distinction of being the most bombed country, in the world.

So now there is just jungles, jungles and jungles with a roads in the middle, small villages around them. The most popular entertainment for a tourists is driving down the river in the wheel of a truck for a few days where locals are throwing a rope to catch for you and after they pull you in for a booze, mushrooms and weed. I passed this fun this time.

I guess it's a very poor country but they have their own sapphire mines and in the few cities cars are new, big and looks like might be quite expensive: like Toyota Land Cruisers and Mitsubishi Pajero. And they are new not the second market! Mostly all homes are built like uplifted from the ground in about two meters with a roofs from bamboos and palm leaves. It reminds me a Tibet when I look into a faces of people but Nepal when I look at surroundings - mountains, plants and green all over around. Just no waterfalls - a pride of Nepal and Himalayan mountains. And instead of rice which is so popular in the rest of Asia people here mostly are cultivating wheat.

I'm moving slowly towards China border crossing point in Bo Ten (Laos) / Meng La (China) which is the only border currently opened to the foreign individual tourists between Laos and China. The distance I've traveled so far is not big it's just that there are mountains all over so the average speed is about 40 km/h and sometimes even that seems much.

The border to China is opened every day from 9 am till 5 pm and my possibilities to make the crossing today where zero even if since morning 8 I had to make only about 170 kilometers. The bus stopped about 30 km before the border right next to the guest house where, of course, tourists are expected to pop-in for a double priced room and now service where offered going further until the border today. But as I'm generally against such a cheap tricks I continued my way towards border hitchhiking which took me about one and a half hours.

So I stayed overnight at some place just few kilometers before the border for 50,000 Kips (which is about five euros) so tomorrow will hopefully enter China.

This border crossing point is only few hundreds kilometers from Tibet which means it will take for me few more days to get till Beijing where my trip officially and finally will be ended.

I woke up before 8 to be ready to wait for a money exchange office right at the border which opens at 8:30 as nobody is accepting Laos money except Laos, probably because for a hundred U.S. dollars you can get a briefcase of Laos Kips. By the way - I was a millionaire in Laos!

I just spent my million to quickly "as my expenses where to high" (paying 5k for a can of a soda) or governmental "bribe" of 300,000.00 Kips to get a visa for entering in Laos...

Entering China went smoothly as I already had made my Chinese visa two months ago while in Kathmandu, Nepal.

And such a relief! Feels the same as coming to China from Russia: from nothing to everything. Even free wi-fi right at the border. Excellent!

So now I'm going in my last journey - crossing all China: from South-West to North-East. And that's it!

Can't believe - I'm home. Only today. Only now. Finally!!! Yeah... 97 days with a backpack, constant moving, hourly change of situations, people and places. Last two days I have spent in the train so no charger and no pictures from that.

I arrived by train from Laos because the hitchhiking is not so welcome in here. I am planning to stay in China and set up some business here. It is much colder in here. But for the next few days I will be relaxing a little bit and probably going to Shanghai.

When I got back home i did not left my bed for a few days. Was just smoking pot and watching movies. More than a few days actually.

Cannot express in words the feeling I get just by being back in China! It is the only place I truly Love on Earth. Just being here makes my heart sing and turns my demons into poets.

Westernized Singapore, beach Thailand, famous U.S., or perfect EU is NOTHING!!! Comparing to Incredibly Amazing-full and Beautiful China with its Imperial Majesty and Complete Harmony of Life. It is something like "all in one package" on every made step.

This is The Only Place I can call HOME, the only place I want to live, the only place I feel naturally happy, the only place I'm really longing for being abroad, the only place i want to return, the only place I Love, the only country I suggest for my true friends.

My Dear China: I'm so happy I'm back and so sorry for wasting my time being abroad for so long and so willing to gain back lost while I was away!!! The feeling is like time used being abroad is the time wasted of my life, of my happiness. I never want to leave you for so long ever again.

And so I'm back! Good morning, China, good morning my dear business city Shanghai, incredible Beijing, posh Shenzhen, unbelievable Guangzhou, so beautiful Guilin, green Fujian, peaceful Tianjin, righteous Lhasa, ice-cold Harbin!!!

Good morning my Home, sweet Home!!!

I'm back and want to stay here forever.

China... Mm ... I really Love You!!!

Can't be amazed by the beautiful buildings and cleanness.

Shanghai. This is what I call a city.

I decided I will stay in China. First of all I have some things to do in here. Secondly, I wish to live here and thirdly I really love china. As I am at Ingus apartment for now he has managed to give to me a separate room - this time I will not sleep on coach as I told him I wish to stay here for longer.

I decided to set up some small internet business- representative fo one company from Ukraine. All these days I was busy on writing business plan, running around the city and looking for deals I need o have before setting all up.

The good thing about China is - you get noticed fast. After my trip I decided to take upon winter swimming. So local Chinese newspaper even published my pictures on this topic.

Also I invited the girl from Thailand which I met on the Bus towards Laos to China to become my wife. She agreed. When we met I didn't had any money at all. I was hungry for days and not properly slept. I was literally homeless. She said yes and later on become my wife.

Checkout how I reached Mt. Kailash at Vedatrac 2 - 108 Gems of Asia

Thank you! ;)

MARTINS ATE'S
108 GEMS OF PURE
vegetarian food

ENJOY YOUR FOOD WITH A PROPER REGGAE CHILLOUT DUB PACKED WITH POSITIVE PRAYERS AND MINDSETS BY zHustlers

zHustlers.com

Made in the USA
Columbia, SC
19 June 2021